STUDENT SOLUTIONS MANUAL

to accompany

DIFFERENTIAL EQUATIONS

GRAPHICS • MODELS • DATA

David Lomen
David Lovelock

Department of Mathematics
University of Arizona

JOHN WILEY & SONS, INC.
New York • Chichester • Weinheim • Brisbane • Singapore • Toronto

COVER ILLUSTRATION Roy Wiemann

ISBN 0-471-32759-X

10 9 8 7 6 5 4 3 2 1

Contents

STUDENT SOLUTIONS MANUAL

to accompany

DIFFERENTIAL EQUATIONS

GRAPHICS • MODELS • DATA

---

1. BASIC CONCEPTS

1.1 Simple Differential Equations and Explicit Solutions

1. **(a)** $y(x) = \int x^3\,dx = \frac{1}{4}x^4 + C$.
$y(1) = 1$, so $1 = \frac{1}{4} + C$, $C = \frac{3}{4}$, and $y(x) = \frac{1}{4}x^4 + \frac{3}{4}$.
See Figure 1-1.

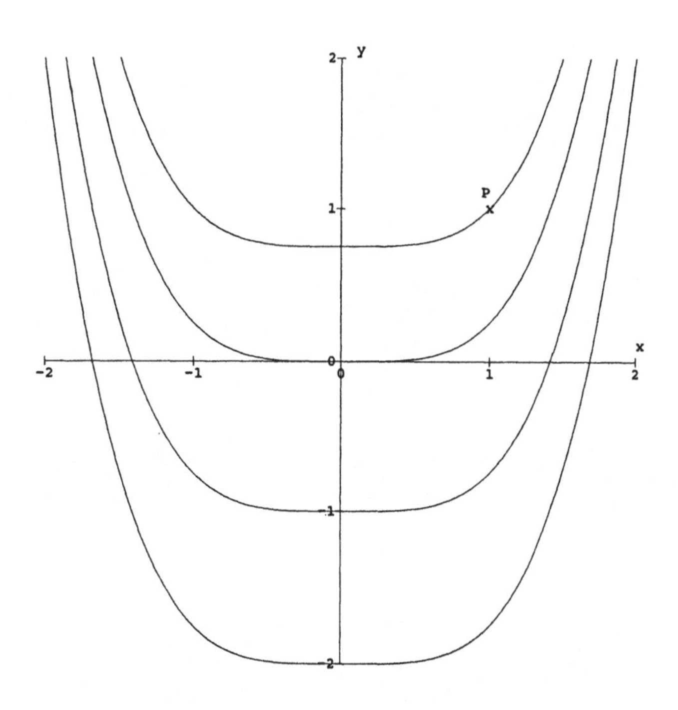

Figure 1-1 Exercise 1(a).

(c) $y(x) = \int \cos x\,dx = \sin x + C$.
$y(0) = 0$, so $0 = 0 + C$; $C = 0$, and $y(x) = \sin x$.
See Figure 1-2.

(e) $y(x) = \int e^{-x}\,dx = -e^{-x} + C$.
$y(0) = 1$, so $1 = -1 + C$; $C = 2$, and $y(x) = -e^{-x} + 2$.
See Figure 1-3.

(g) $y(x) = \int \frac{1}{x}\,dx = \ln|x| + C$.
$y(-1) = 1$, so $1 = 0 + C$; $C = 1$, and $y(x) = \ln|x| + 1$, $x < 0$. $y(x) = \ln(-x) + 1$.
Note: An answer of $y(x) = \ln|x| + 1$ is not correct.
See Figure 1-4.

(i) Using partial fractions we have

$$\frac{1}{x(1-x)} = \frac{A}{x} + \frac{B}{1-x} = \frac{A(1-x) + Bx}{x(1-x)} = \frac{(B-A)x + A}{x(1-x)}.$$

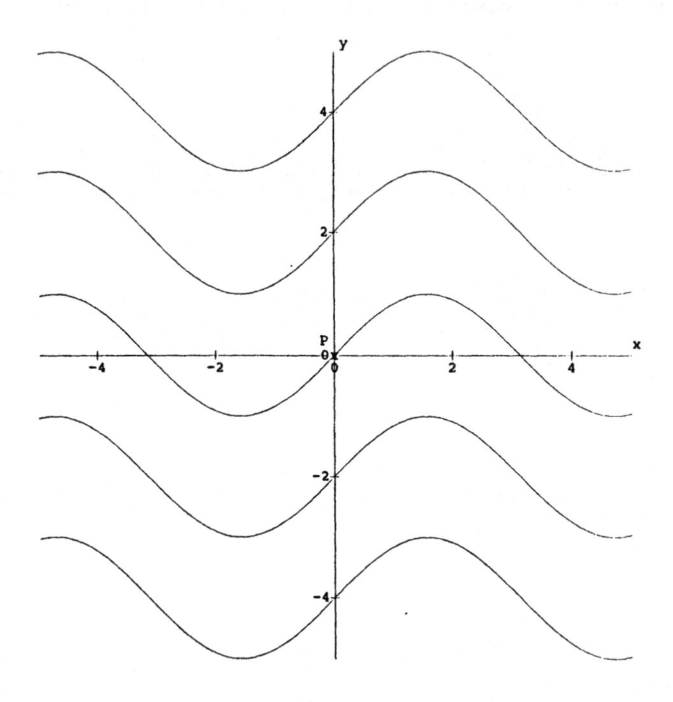

Figure 1-2 Exercise 1(c).

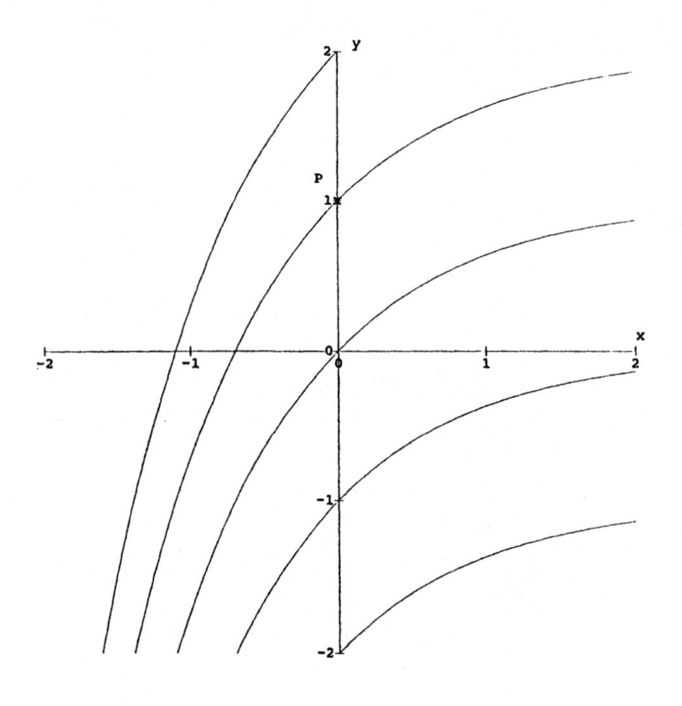

Figure 1-3 Exercise 1(e).

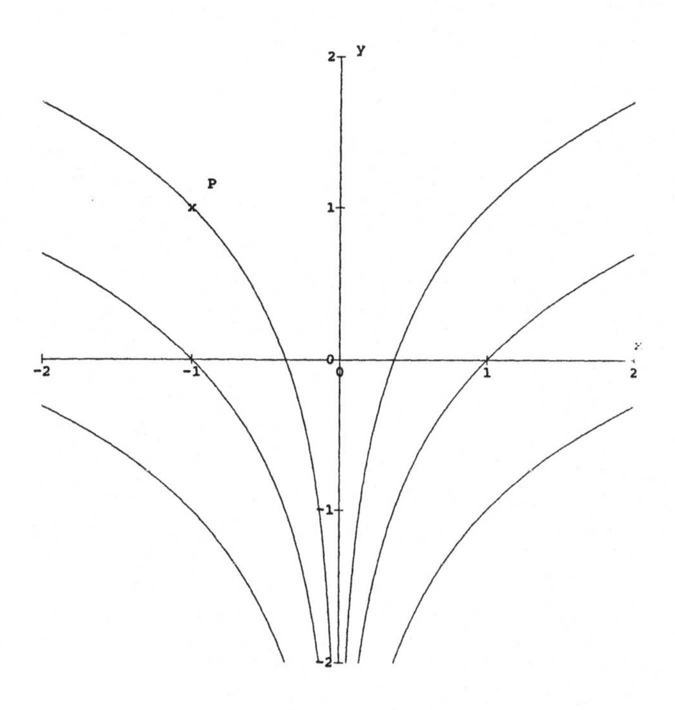

Figure 1-4 Exercise 1(g).

This gives $1 = (B - A)x + A$, or $1 = A$, $0 = B - A$, so $A = 1$, $B = 1$, and

$$\frac{1}{x(1-x)} = \frac{1}{x} + \frac{1}{1-x}.$$

$$y(x) = \int \frac{1}{x(1-x)}\,dx = \int \frac{1}{x}\,dx + \int \frac{1}{1-x}\,dx = \ln|x| - \ln|1-x| + C.$$

$y(2) = 1$, so $1 = \ln 2 + C$, $C = 1 - \ln 2$, and

$$y(x) = \ln x - \ln(x-1) + 1 - \ln 2 = \ln\left[\frac{x}{2(x-1)}\right] + 1,\, x > 1.$$

See Figure 1-5.

(k) $y(x) = \int x^2 e^{-x}\,dx = -x^2 e^{-x} - 2xe^{-x} - 2e^{-x} + C.$
$y(0) = 1$, so $1 = -2 + C$; $C = 3$; $y(x) = -x^2 e^{-x} - 2xe^{-x} - 2e^{-x} + 3.$
See Figure 1-6.

3. $S(x) = \sqrt{\frac{2}{\pi}} \int_0^x \sin t^2\,dt.$

(a) $S(0) = \sqrt{\frac{2}{\pi}} \int_0^0 \sin t^2\,dt = \sqrt{\frac{2}{\pi}} \cdot 0 = 0.$

(b) $S(-x) = \sqrt{\frac{2}{\pi}} \int_0^{-x} \sin t^2\,dt$, so we change the dummy variable of integration by $t = -z$,
$S(-x) = -\sqrt{\frac{2}{\pi}} \int_0^x \sin z^2\,dz = -S(x)$. Thus, $S(x)$ is an odd function of x.

(c)-(e) Using Simpson's rule we can construct the following table.

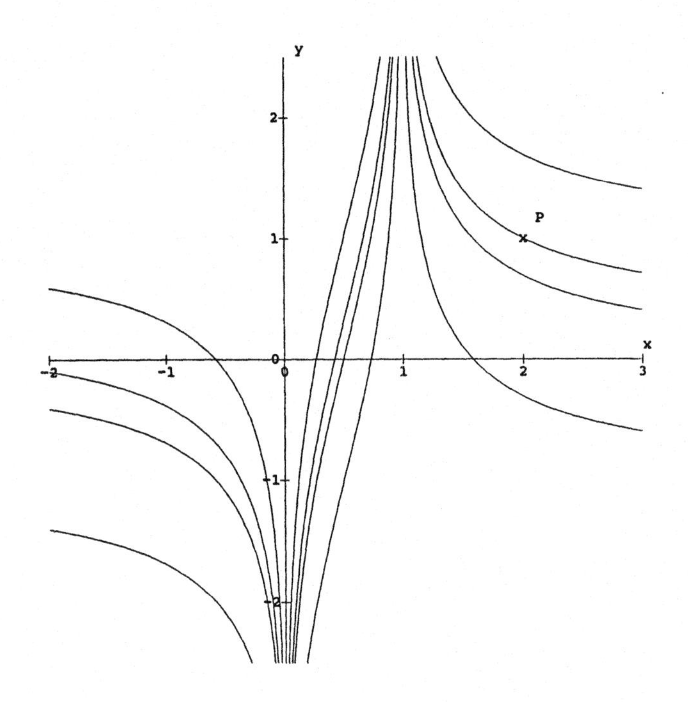

Figure 1-5 Exercise 1(i).

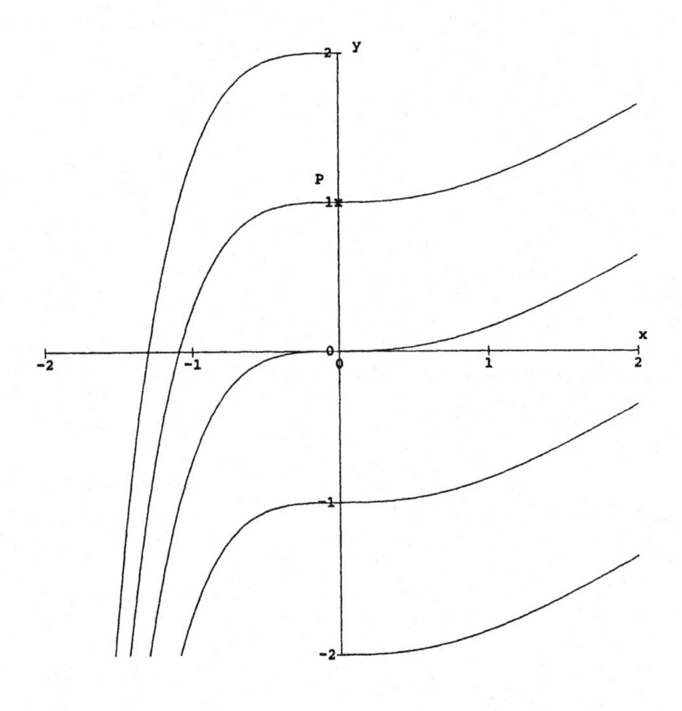

Figure 1-6 Exercise 1(k).

x	$y(x)$	x	$y(x)$
0.5	0.033	3.0	0.617
1.0	0.248	3.5	0.394
1.5	0.621	4.0	0.596
2.0	0.642	4.5	0.483
2.5	0.344	5.0	0.421

Notice that if we plot the values of y corresponding to $x = 2$ and 4 we appear to have a decreasing function.

The table, together with a possible plot of $S(x)$, are shown in Figure 1-7.

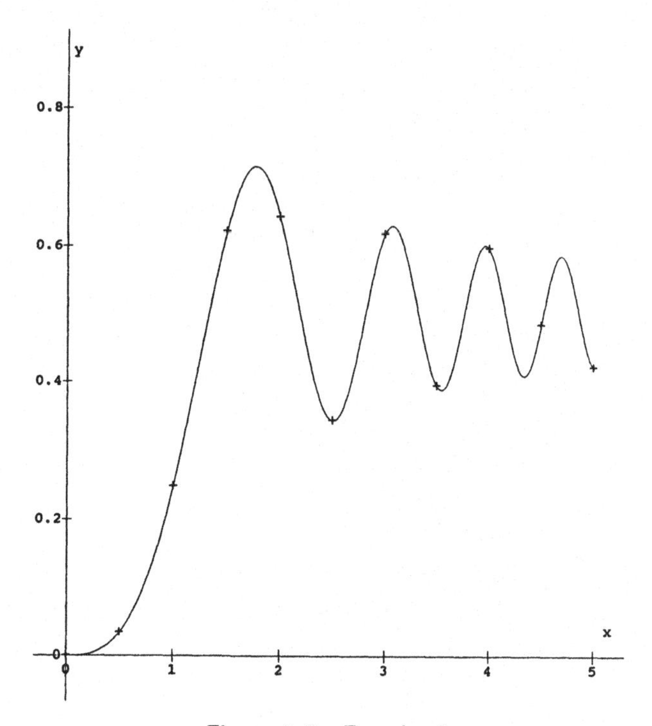

Figure 1-7 Exercise 3.

(f) Using Simpson's rule we find $S(50) \approx 0.494$, $S(100) \approx 0.504$, $S(150) \approx 0.497$, $S(200) \approx 0.499$, $S(250) \approx 0.499$. This suggests that as $x \to \infty$, $S(x) \to 0.5$.

5. If $y' = e^x$ then $y(x) = e^x + C_1$. If $y' = -e^{-x}$ then $y(x) = e^{-x} + C_2$.

The angle of intersection between two curves is the angle of intersection between their tangent lines. Here the angle of intersection seems to be a right angle or 90°. If the two differential equations are multiplied together, the result is -1. Since the differential equations represent the slopes of the families of solutions, the fact that their product equals -1 informs us that they meet at a right angle. Another way of saying this is perpendicular slopes are negative reciprocals of one another. Here $y' = e^x$ is the negative reciprocal of $y' = -e^{-x}$.

1.2 Graphical Solutions Using Calculus

1. For Exercises 1(a) through (l): When we are sketching various solution curves for the first order differential equations, uniqueness assures us that the curves cannot cross.

(a) $y' = x^3$, $y'' = 3x^2$.

Monotonicity: The solutions are increasing ($y' > 0$) when $x > 0$ and decreasing ($y' < 0$) when $x < 0$. At $x = 0$, $y' = 0$ indicates a horizontal tangent line.

Concavity: The solutions are concave up ($y'' > 0$) for all x, except for $x = 0$ where $y'' = 0$. Therefore, the solutions have a minimum at $x = 0$.

Symmetry: In $\frac{dy}{dx} = x^3$ we replace x by $-x$ to find $\frac{dy}{-dx} = (-x)^3$, that is $\frac{dy}{dx} = x^3$, which is the original differential equation. This shows that the family of solutions is symmetric across the y-axis. In $\frac{dy}{dx} = x^3$ we replace x by $-x$ and y by $-y$ to find $\frac{-dy}{-dx} = (-x)^3$, that is $\frac{dy}{dx} = -x^3$, which is not the original differential equation. The family of solutions is not symmetric about the origin.

Singularities: There are no obvious points where the derivative fails to exist.

See Figure 1-1.

(c) $y' = \cos x$, $y'' = -\sin x$.

Monotonicity: The solutions are increasing ($y' > 0$) when $-\frac{\pi}{2} < x < \frac{\pi}{2}$, $\frac{3\pi}{2} < x < \frac{5\pi}{2}$, and so on — that is, $(4n-1)\frac{\pi}{2} < x < (4n+1)\frac{\pi}{2}$ for $n = 0, \pm 1, \pm 2, \cdots$. The solutions are decreasing ($y' < 0$) when $-\frac{3\pi}{2} < x < -\frac{\pi}{2}$, $\frac{\pi}{2} < x < \frac{3\pi}{2}$, and so on — that is, $(4n+1)\frac{\pi}{2} < x < (4n+3)\frac{\pi}{2}$ for $n = 0, \pm 1, \pm 2, \cdots$.

Concavity: The solutions are concave up ($y'' > 0$) when $-\pi < x < 0$, $\pi < x < 2\pi$, and so on — that is, $(2n-1)\pi < x < 2n\pi$ for $n = 0, \pm 1, \pm 2, \cdots$. The solutions are concave down ($y'' < 0$) when $-2\pi < x < -\pi$, $0 < x < \pi$, and so on — that is, $2n\pi < x < (2n+1)\pi$ for $n = 0, \pm 1, \pm 2, \cdots$. The solutions have inflection points ($y'' = 0$) when $x = -2\pi, -\pi, 0, \pi, 2\pi$ and so on — that is, $x = n\pi$ for $n = 0, \pm 1, \pm 2, \cdots$.

Symmetry: In $\frac{dy}{dx} = \cos x$ we replace x by $-x$ to find $\frac{dy}{-dx} = \cos(-x)$, that is $\frac{dy}{dx} = -\cos x$, which is not the original differential equation. This shows that the family of solutions is not symmetric across the y-axis. In $\frac{dy}{dx} = \cos x$ we replace x by $-x$ and y by $-y$ to find $\frac{-dy}{-dx} = \cos(-x)$, that is $\frac{dy}{dx} = \cos x$, which is the original differential equation. The family of solutions is symmetric about the origin.

Singularities: There are no obvious points where the derivative fails to exist.

See Figure 1-2.

(e) $y' = e^{-x}$, $y'' = -e^{-x}$.

Monotonicity: The solutions are increasing ($y' > 0$) for all x.

Concavity: The solutions are concave down ($y'' < 0$) for all x.

Symmetry: In $\frac{dy}{dx} = e^{-x}$ we replace x by $-x$ to find $\frac{dy}{-dx} = e^{x}$, that is $\frac{dy}{dx} = -e^{x}$, which is not the original differential equation. This shows that the family of solutions is not symmetric across the y-axis. In $\frac{dy}{dx} = e^{-x}$ we replace x by $-x$ and y by $-y$ to find $\frac{-dy}{-dx} = e^{x}$, that is $\frac{dy}{dx} = e^{x}$, which is not the original differential equation. The family of solutions is not symmetric about the origin.

Singularities: There are no obvious points where the derivative fails to exist.

See Figure 1-3.

(g) $y' = \frac{1}{x}$, $y'' = -\frac{1}{x^2}$.

Monotonicity: The solutions are increasing ($y' > 0$) when $x > 0$ and decreasing when $x < 0$.

Concavity: The solutions are concave down ($y'' < 0$) for all x, except for $x = 0$ where y'' is undefined.

Symmetry: The family of solutions is symmetric across the y-axis. However, the family of solutions is not symmetric about the origin.

Singularities: Because the derivative is undefined at $x = 0$, we anticipate problems at $x = 0$.

See Figure 1-4.

(i) $y' = 1/[x(1-x)]$ $y'' = (-1+2x)/[x(1-x)]^2$

Monotonicity: The solutions are increasing ($y' > 0$) when $x(1-x) > 0$, that is, $0 < x < 1$. The solutions are decreasing ($y' < 0$) when $x(1-x) < 0$, that is, $x < 0$ or $1 < x$.

Concavity: Because $[x(1-x)]^2 \geq 0$ for all x the solutions are concave up when $x > 1/2$ and concave down when $x < 1/2$. Inflection point $x = 1/2$.

Symmetry: The family of solutions is not symmetric across the x-axis or about the origin.

See Figure 1-5.

Singularities: Because the derivative is not defined at $x = 0$ and $x = 1$, we anticipate problems at these two values of x.

(k) $y' = x^2e^{-x}$, $y'' = 2xe^{-x} - x^2e^{-x} = x(2-x)e^{-x}$.

Monotonicity: The solutions are increasing ($y' > 0$) for all x except $x = 0$ where the tangent lines will be horizontal.

Concavity: The solutions are concave up ($y'' > 0$) when $0 < x < 2$ and concave down ($y'' < 0$) when $x < 0$ or $x > 2$.

Symmetry: The family of solutions is not symmetric across the y-axis or about the origin.

Singularities: There are no obvious points where the derivative fails to exist.

See Figure 1-6.

3. Using partial fractions we have

$$\frac{4}{x(x-4)} = \frac{A}{x} + \frac{B}{x-4} = \frac{A(x-4)+Bx}{x(x-4)} = \frac{(A+B)x-4A}{x(x-4)}.$$

This gives $4 = (A+B)x - 4A$, or $4 = -4A$, $0 = A+B$, so $A = -1$, $B = 1$, and $4/[x(x-4)] = -1/x + 1/(x-4)$. Integration gives

$$\begin{aligned}y(x) &= \int 4/[x(x-4)]\,dx \\ &= -\int 1/x\,dx + \int 1/(x-4)\,dx \\ &= -\ln|x| + \ln|x-4| + C \\ &= \ln\left|\tfrac{x-4}{x}\right| + C.\end{aligned}$$

(a) $y(-1) = 0$, so $0 = \ln 5 + C$, giving $C = -\ln 5$ and $y(x) = \ln[(x-4)/(5x)]$, $x < 0$.

(b) $y(1) = 0$, so $0 = \ln 3 + C$, giving $C = -\ln 3$ and $y(x) = \ln[(4-x)/(3x)]$, $0 < x < 4$.

(c) $y(5) = 0$, so $0 = \ln(1/5) + C$, giving $C = \ln 5$ and $y(x) = \ln[5(x-4)/x]$, $x > 4$.

5. $y' = \sqrt{\frac{2}{\pi}}\sin x^2$, $y'' = 2\sqrt{\frac{2}{\pi}}x\cos x^2$.

Monotonicity: Horizontal tangent lines will occur when $\sin x^2 = 0$, that is, when $x = \pm\sqrt{n\pi}$, for $n = 0, 1, 2, \cdots$. The solutions are increasing ($y' > 0$) when $\sin x^2 > 0$ — that is, when $2n\pi < x^2 < (2n+1)\pi$, $n = 0, 1, 2, \cdots$. Thus, solutions are increasing when $\sqrt{2n\pi} < x < \sqrt{(2n+1)\pi}$, $n = 0, 1, 2, \cdots$. In the same way, solutions are decreasing when $\sqrt{(2n+1)\pi} < x < \sqrt{(2n+2)\pi}$, $n = 0, 1, 2, \cdots$.

n	$\sqrt{n\pi}$	n	$\sqrt{n\pi}$
1	1.772	6	4.342
2	2.507	7	4.689
3	3.070	8	5.013
4	3.545	9	5.319
5	3.963	10	5.605

Concavity: Possible points of inflection occur when $x \cos x^2 = 0$, that is, when $x = 0$ or $x = \pm\sqrt{\pi/2 + n\pi}$, for $n = 0, 1, 2, \cdots$. The solutions are concave up ($y'' > 0$) when $x \cos x^2 > 0$. For $x > 0$, this occurs when $-\pi/2 + 2n\pi < x^2 < \pi/2 + 2n\pi$, n any nonnegative integer. This tells us that the solution curves are concave up when $0 < x < \pi/2$ or $\sqrt{-\pi/2 + 2n\pi} < x < \sqrt{\pi/2 + 2n\pi}$, $n = 1, 2, 3, \cdots$. In the same way, for $x > 0$, the solution curves will be concave down whenever $\cos x^2 < 0$, that is, when $\sqrt{\pi/2 + 2n\pi} < x < \sqrt{3\pi/2 + 2n\pi}$, $n = 0, 1, 2, \cdots$.

Symmetry: The family of solutions is not symmetric across the y-axis. However, the family of solutions is symmetric about the origin.

Singularities: There are no obvious points where the derivative fails to exist.

See Figure 1-7. As $x \to \infty$, $y(x) \to 0.5$.

7. The statement that two solutions intersect means that there is a common point (x_0, y_0) through which two distinct particular solutions of $y' = g(x)$, say $y_1(x)$ and $y_2(x)$, pass. Because both y_1 and y_2 are solutions of $y' = g(x)$, we must have $y_1' = g(x)$ and $y_2' = g(x)$, so that $y_1' = y_2'$, or $(y_1 - y_2)' = 0$. From this we have $y_1(x) - y_2(x) = C$. The fact that $y_0 = y_1(x_0)$ and $y_0 = y_2(x_0)$ requires that $C = 0$, so that $y_1(x) = y_2(x)$. In other words the two curves $y_1(x)$ and $y_2(x)$ are one and the same. This means that only one solution of $y' = g(x)$ can pass through any point (x_0, y_0).

1.3 Slope Fields and Isoclines

1. Compare your answers with those you found for Exercise 1, Section 1.1 and Exercise 1, Section 1.2.

(a) The slope field and some solution curves are shown in Figure 1-8.

Isoclines are given by $x^3 = m$, that is, $x = m^{1/3}$, so isoclines are defined for every m. The isocline for slope $m = 0$ is the vertical line $x = 0$, and this agrees with the slope field shown in Figure 1-8, where we have also shown isoclines of slope $m = 1$ and $m = -1$.

(c) The slope field and some solution curves are shown in Figure 1-9.

Isoclines are given by $\cos x = m$, that is, $x = \pm \arccos m + 2n\pi$, $n = 0, \pm1, \pm2, \cdots$, so isoclines are defined for $-1 \leq m \leq 1$. The isocline for slope $m = 0$ consists of the vertical lines $x = \pm\frac{\pi}{2} + 2n\pi$, $n = 0, \pm1, \pm2, \cdots$, and this agrees with the slope field shown in Figure 1-9, where we have also shown the isocline of slope $m = -1$.

(e) The slope field and some solution curves are shown in Figure 1-10.

Isoclines are given by $e^{-x} = m$, that is, $x = \ln\frac{1}{m}$, so isoclines are defined for $m > 0$. The isocline for slope $m = 1$ is the vertical line $x = 0$, and this agrees with the slope field shown in Figure 1-10, where we have also shown isoclines of slope $m = 2$ and $m = \frac{1}{2}$.

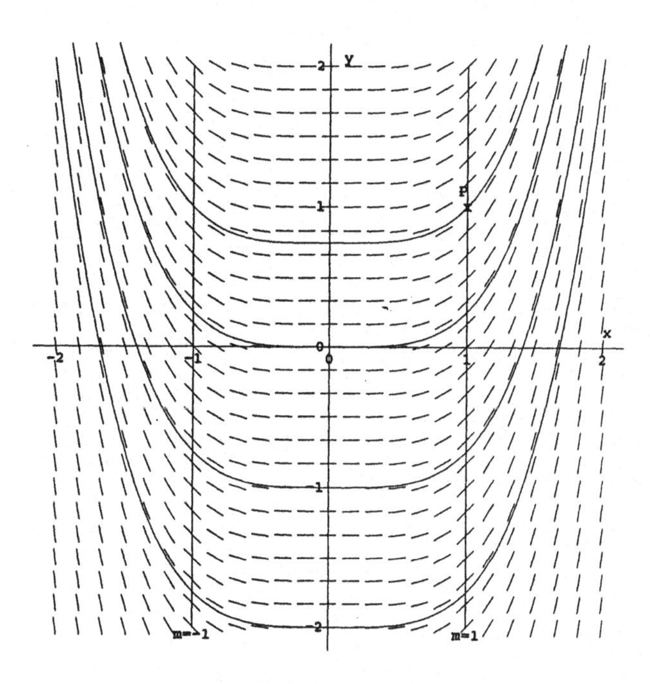

Figure 1-8 Exercise 1(a).

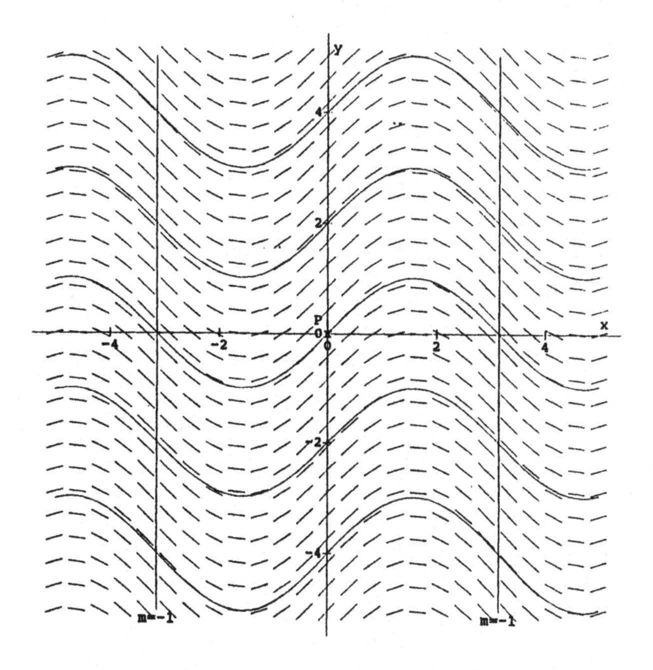

Figure 1-9 Exercise 1(c).

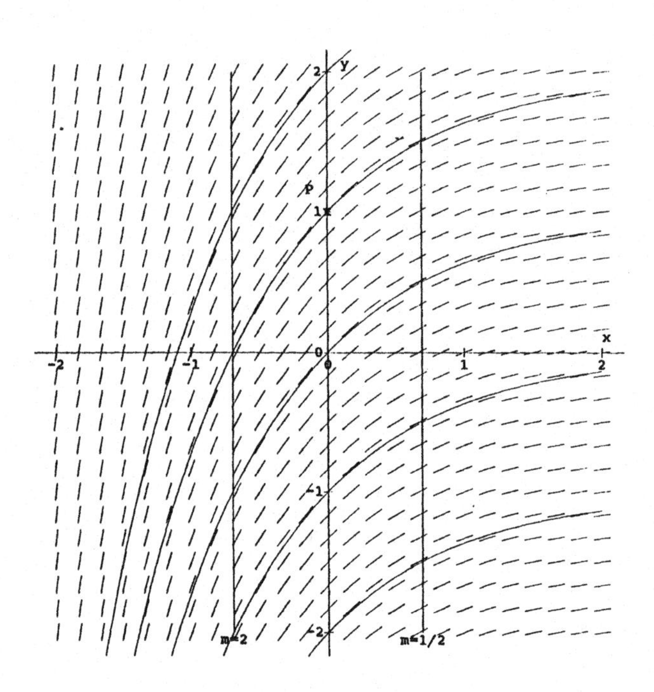

Figure 1-10 Exercise 1(e).

(g) The slope field and some solution curves are shown in Figure 1-11.

Isoclines are given by $\frac{1}{x} = m$, that is, $x = \frac{1}{m}$, so isoclines are defined for every $m \neq 0$. The isocline for slope $m = 1$ is the vertical line $x = 1$, and this agrees with the slope field shown in Figure 1-11, where we have also shown the isocline of slope $m = -1$.

(i) The slope field and some solution curves are shown in Figure 1-12.

Isoclines are given by $\frac{1}{x(1-x)} = m$, that is, $x = \frac{1}{2} \pm \sqrt{\frac{m-4}{4m}}$, so isoclines are defined for $m \geq 4$ and $m < 0$. The isocline for slope $m = 4$ is the vertical line $x = 1/2$, and this agrees with the slope field shown in Figure 1-12, where we have also shown the isocline of slope $m = -1$.

(k) The slope field and some solution curves are shown in Figure 1-13.

Isoclines are given by $x^2 e^{-x} = m$. It is not possible to solve this explicitly for x, but we see that isoclines are defined for $m \geq 0$. We can still identify isoclines by choosing any x, and substituting this into $x^2 e^{-x} = m$ to find the corresponding m. Thus, the vertical line isocline $x = 0$ corresponds to slope $m = 0$, the vertical line isocline $x = 1$ corresponds to slope $m = 1/e$, and the vertical line isocline $x = -1$ corresponds to slope $m = e$. This agrees with the slope field shown in Figure 1-13.

3. Compare your answers with those you found for Exercise 3, Section 1.1 and Exercise 3, Section 1.2.

The slope field and some solution curves are shown in Figure 1-14.

Isoclines are given by $\sqrt{\frac{2}{\pi}} \sin x^2 = m$, that is $x = \pm\sqrt{\arcsin\left(m\sqrt{\frac{\pi}{2}}\right) + 2n\pi}$ and $x = \pm\sqrt{\pi - \arcsin\left(m\sqrt{\frac{\pi}{2}}\right) + 2n\pi}$, $n = 0, \pm 1, \pm 2, \cdots$, so isoclines are defined for $-1 \leq m\sqrt{\frac{\pi}{2}} \leq 1$, that is, $-\sqrt{\frac{2}{\pi}} \leq m \leq \sqrt{\frac{2}{\pi}}$.As $x \to \infty$, if $y(0) = 0$, then $y(x) \to \frac{1}{2}$.

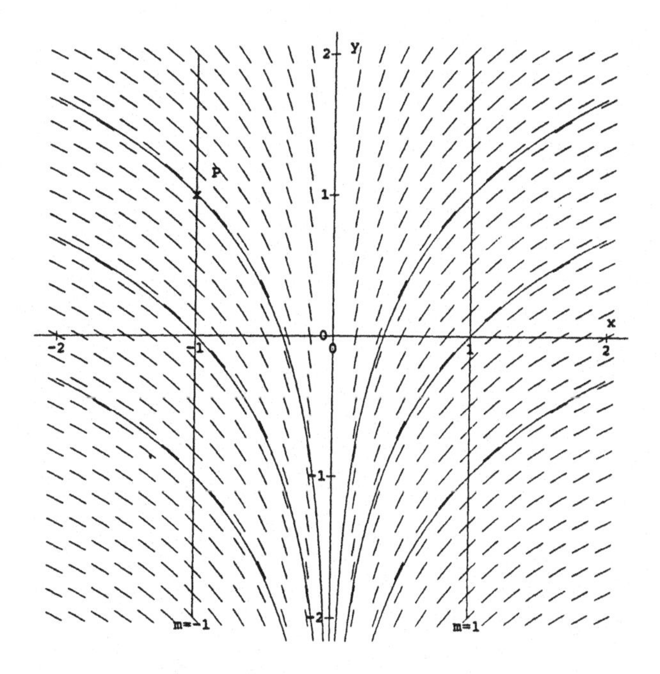

Figure 1-11 Exercise 1(g).

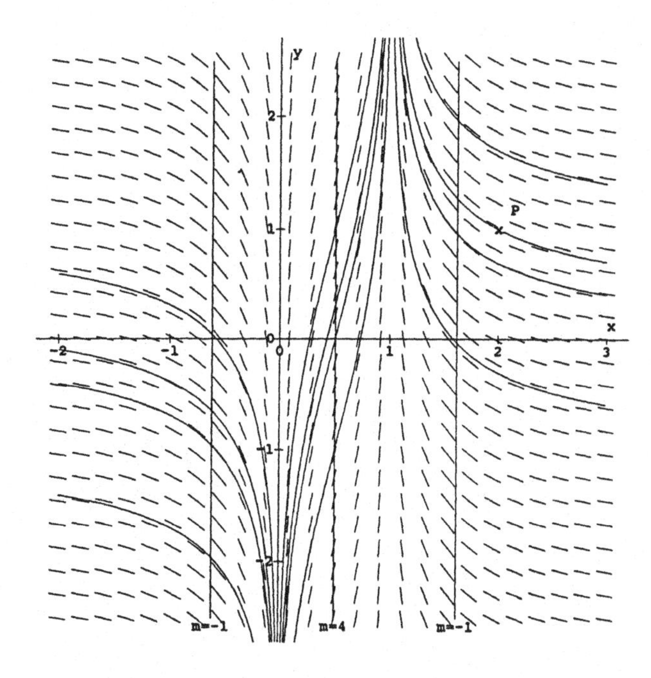

Figure 1-12 Exercise 1(i).

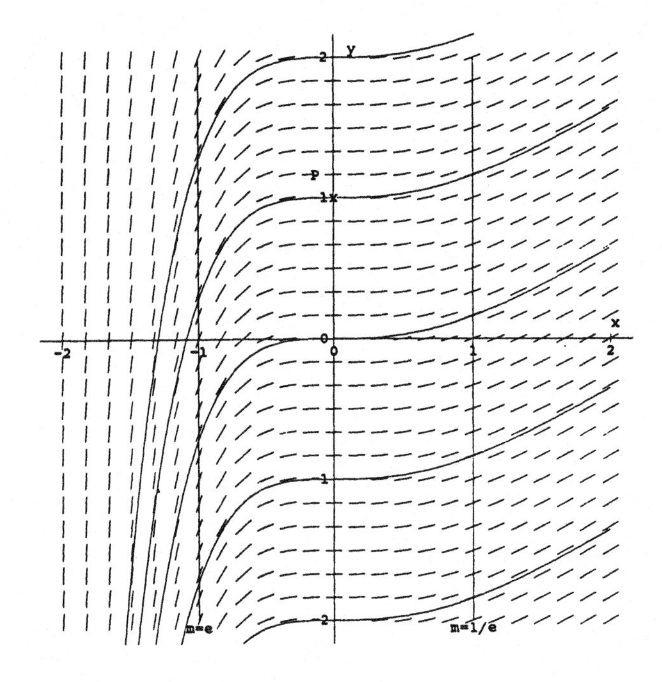

Figure 1-13 Exercise 1(k).

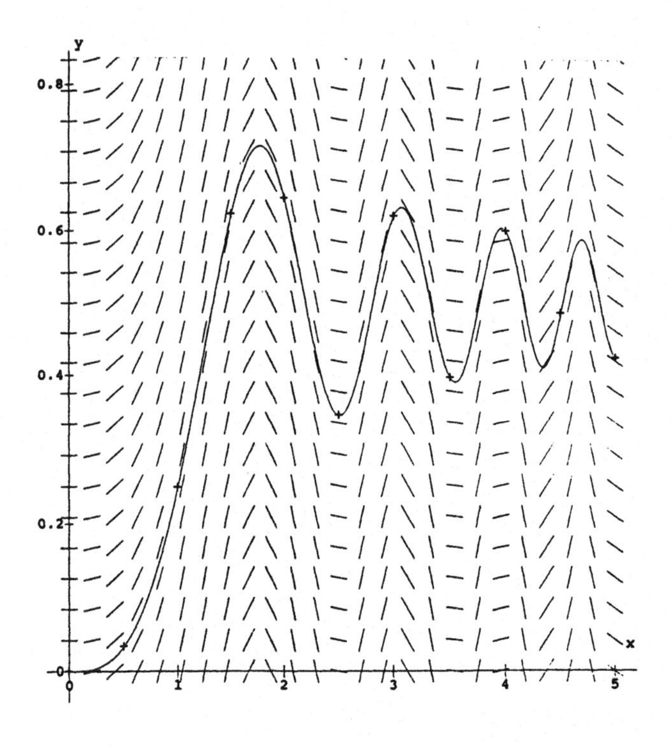

Figure 1-14 Exercise 3.

5. **(a)** If $y(x)$ is one solution of $y' = g(x)$, then $y(x) + C$ is another solution for any value of the constant C. Thus, other solutions can be plotted from the given solution by translating the given solution vertically.

(b) No. The given solution might have a vertical asymptote at $x = -1$, or to the left of $x = -1$. If it has a vertical asymptote at $x = -1$, then any solution for $x < -1$ cannot be obtained from the given solution by a vertical translation.

7. **(a)** The slope field given in Figure 1.20 of the text is $y' = \ln|x|$.

The solutions of $y' = x + 1$ are decreasing for $x < -1$, increasing for $x > -1$, and have horizontal tangent lines only when $x = -1$, which eliminates $y' = x + 1$ as a possibility.

The solutions of $y' = x - 1$ are decreasing for $x < 1$ and increasing for $x > 1$, and have horizontal tangent lines only when $x = 1$, which eliminates $y' = x - 1$ as a possibility.

The solutions of the remaining two differential equations, $y' = \ln|x|$ and $y' = x^2 - 1$ share the following characteristics: increasing for $x < -1$ or $x > 1$, decreasing for $-1 < x < 1$, concave up for $x > 0$, concave down for $x < 0$, and symmetric about the origin. The main difference is found when looking at the singularities of these two differential equations. $y' = \ln|x|$ is undefined for $x = 0$. Therefore, the solutions of $y' = \ln|x|$ will have a vertical asymptote at $x = 0$, while the solutions of $y' = x^2 - 1$ will have slopes near -1 close to the y-axis.

(b) A general strategy for matching differential equations with a given graph is:

1. Look at the graph and determine where y is increasing and decreasing (monotonicity).

2. Look at the graph and identify where the graph is concave up and concave down(concavity).

3. Identify any asymptotes that the graph has.

4. Eliminate all differential equations that do not fit these conditions.

5. Look at singularities and symmetries in the remaining differential equations. Eliminate all differential equations that do not fit these conditions.

9. Note that $y(0)$ represents the initial position rather than the distance above the ground.

(a) From $y' = 49(1 - e^{-0.2x})$ we see that $y' > 0$ if $x > 0$ and $y' < 0$ if $x < 0$. Also, from $y'' = 49 \times 0.2e^{-0.2x}$ we see that $y'' > 0$ for all $x > 0$. Thus, the solutions of the differential equation are increasing for $x > 0$, decreasing for $x < 0$, and concave up for all x. The slope field is shown in Figure 1-15, where we have drawn the solution with $y(0) = 0$ by hand. We notice that after a very short time the slope field appears to have a constant slope of 49. Thus, we can estimate when a line with slope 49 starting from $(0, 0)$ will reach $y = 5000$, which is at $5000/49 \approx 102$ seconds. Because $49(1 - e^{-0.2x}) < 49$ we expect the actual solution to take longer to reach $y = 5000$, so 102 seconds is an underestimate.

(b) $y' = 49(1 - e^{-0.2x})$,

$y(x) = 49\left[x - \left(-\frac{1}{0.2}\right)\left(e^{-0.2x}\right)\right] + C$,

$y(x) = 49x + 245e^{-0.2x} + C$.

Now substitute to find C using the initial condition that was given: $y(0) = 0$.

$0 = (49)(0) + (245)(e^0) + C$,

$C = -245$. So,

$y(x) = 49x + 245e^{-0.2x} - 245$.

Let T be the time that the object takes to hit the ground, so

$y(T) = 5000$ and thus,

$5000 = 49T + 245e^{-0.2T} - 245$,

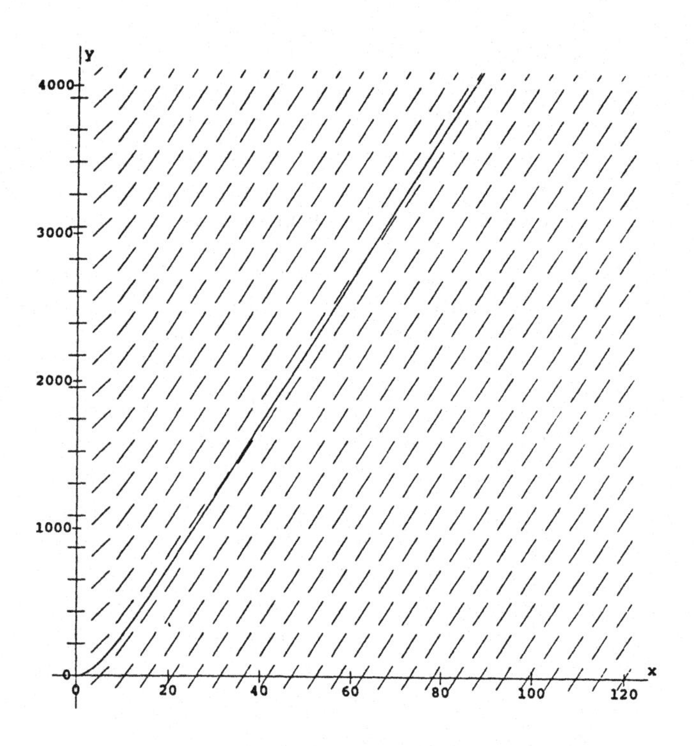

Figure 1-15 Exercise 9(a).

$5245 = 49T + 245e^{-0.2T}$.
Either graphically or numerically solve for T.
$T \approx 107.04$ seconds.

1.4 Functions and Power Series Expansions

1. Recall that the function $\sin x$ has the Taylor series expansion

$$\sin x = \sum_{k=0}^{\infty} (-1)^k \frac{x^{2k+1}}{(2k+1)!} = x - \frac{x^3}{3!} + \frac{x^5}{5!} - \frac{x^7}{7!} + \cdots,$$

valid for all x. If in this equation we replace x by t^2, we find

$$\sin t^2 = \sum_{k=0}^{\infty} (-1)^k \frac{t^{4k+2}}{(2k+1)!} = t^2 - \frac{t^6}{3!} + \frac{t^{10}}{5!} - \frac{t^{14}}{7!} + \cdots,$$

which, when integrated from 0 to x, gives

$$\int_0^x \sin t^2 \, dt = \sum_{k=0}^{\infty} (-1)^k \int_0^x \frac{t^{4k+2}}{(2k+1)!} \, dt = \sum_{k=0}^{\infty} (-1)^k \frac{x^{4k+3}}{(4k+3)(2k+1)!}.$$

Thus, the Fresnel Sine integral can be written in the form

$$\sqrt{\frac{2}{\pi}} \int_0^x \sin t^2 \, dt = \sqrt{\frac{2}{\pi}} \sum_{k=0}^{\infty} (-1)^k \frac{x^{4k+3}}{(4k+3)(2k+1)!} = \sqrt{\frac{2}{\pi}} \left(\frac{x^3}{3 \cdot 1!} - \frac{x^7}{7 \cdot 3!} + \frac{x^{11}}{11 \cdot 5!} - \frac{x^{15}}{15 \cdot 7!} + \cdots \right.$$

valid for all x.

In Figure 1-16 we have plotted $S(x)$ from Figure 1-7, and the four functions

$$P_1(x) = \sqrt{\frac{2}{\pi}}\left(\frac{x^3}{3 \cdot 1!}\right),$$

$$P_2(x) = \sqrt{\frac{2}{\pi}}\left(\frac{x^3}{3 \cdot 1!} - \frac{x^7}{7 \cdot 3!}\right),$$

$$P_3(x) = \sqrt{\frac{2}{\pi}}\left(\frac{x^3}{3 \cdot 1!} - \frac{x^7}{7 \cdot 3!} + \frac{x^{11}}{11 \cdot 5!}\right),$$

$$P_4(x) = \sqrt{\frac{2}{\pi}}\left(\frac{x^3}{3 \cdot 1!} - \frac{x^7}{7 \cdot 3!} + \frac{x^{11}}{11 \cdot 5!} - \frac{x^{15}}{15 \cdot 7!}\right).$$

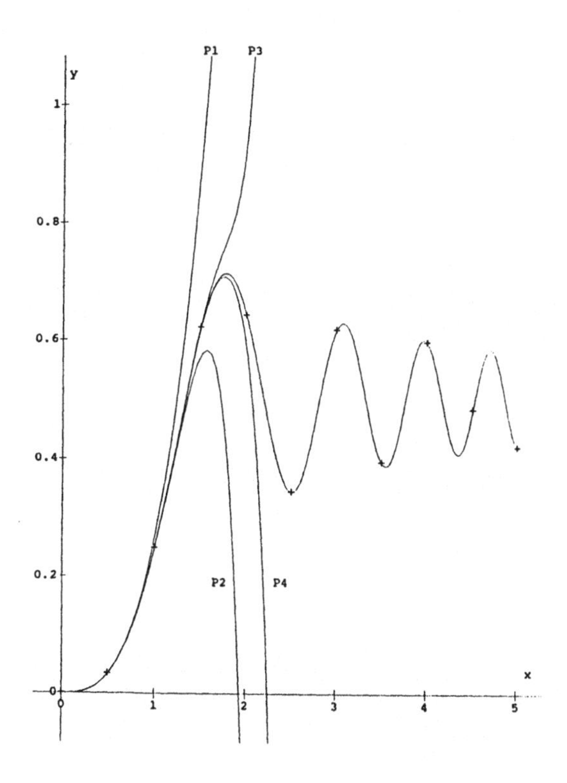

Figure 1-16 Exercise 1.

2. AUTONOMOUS DIFFERENTIAL EQUATIONS

2.1 Autonomous Equations

1. **(a)** i. $y' = -\frac{1}{y^2}$, $y'' = \frac{2}{y^3}y' = -\frac{2}{y^5}$. The solutions are decreasing ($y' < 0$) for all $y \neq 0$, concave up ($y'' > 0$) for $y < 0$, and concave down ($y'' < 0$) for $y > 0$. Isoclines are given by $-\frac{1}{y^2} = m$, that is, $y = \pm\sqrt{-1/m}$. The slope field, the isocline for slope $m = -1$, and some solution curves are shown in Figure 2-1.

ii. There are no equilibrium solutions. $\int y^2\,dy = -\int dx$ gives $\frac{1}{3}y^3 = -x + C$, which can be solved for y: $y(x) = [3(-x + C)]^{1/3}$.

iii. Parts i and ii are consistent.

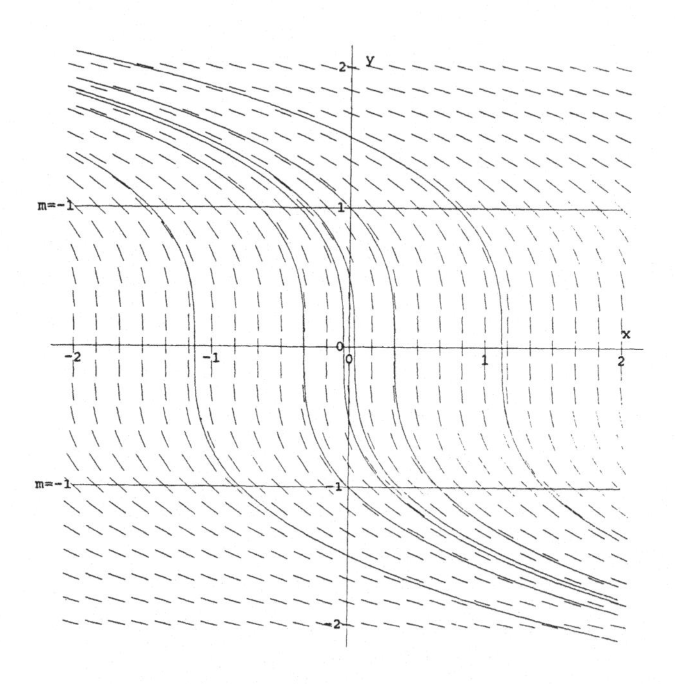

Figure 2-1 Exercise 1(a).

(c) i. $y' = y^2$, $y'' = 2yy' = 2y^3$. The solutions are increasing ($y' > 0$) for all $y \neq 0$, concave up ($y'' > 0$) for $y > 0$, and concave down ($y'' < 0$) for $y < 0$. Isoclines are given by $y^2 = m$, that is, $y = \pm\sqrt{m}$. The slope field, the isocline for slope $m = 1$, and some solution curves are shown in Figure 2-2.

ii. The equilibrium solution is $y(x) = 0$. Nonequilibrium solutions are obtained from $\int \frac{1}{y^2}\,dy = \int dx$, which gives $-\frac{1}{y} = x + C$, or $y(x) = \frac{-1}{x+C}$. Thus, the complete solution consists of two parts: $y(x) = 0$ and $y(x) = \frac{-1}{x+C}$.

iii. Parts i and ii should be consistent, but the analytical solution from part ii, namely, $y(x) = \frac{-1}{x+C}$, shows there are vertical asymptotes at $x = -C$ which are impossible to detect from the analysis in part i.

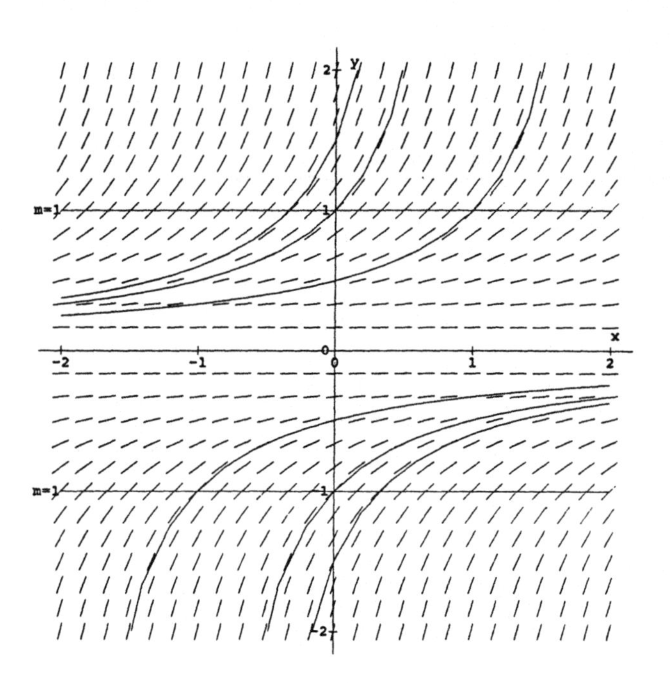

Figure 2-2 Exercise 1(c).

(e) i. $y' = y^{3/2}$, $y'' = \frac{3}{2}y^{1/2}y' = \frac{3}{2}y^2$. The solutions are defined only for $y \geq 0$, because $y^{3/2}$ is undefined for $y < 0$. The solutions are increasing ($y' > 0$) for all $y > 0$, and concave up ($y'' > 0$) for all $y > 0$. Isoclines are given by $y^{3/2} = m$, that is, $y = m^{2/3}$. The slope field, the isoclines for slope $m = 1$ and $m = 2$, and some solution curves are shown in Figure 2-3.

ii. The equilibrium solution is $y(x) = 0$. Nonequilibrium solutions are obtained from $\int \frac{1}{y^{3/2}}\,dy = \int dx$, which gives $-2\frac{1}{y^{1/2}} = x + C$, or $y(x) = \left(-\frac{2}{x+C}\right)^2$ for $x + C < 0$. Thus, the complete solution consists of two parts: $y(x) = 0$ and $y(x) = \left(-\frac{2}{x+C}\right)^2$ for $x + C < 0$.

iii. Parts i and ii should be consistent, but the analytical solution from part ii, namely, $y(x) = \left(-\frac{2}{x+C}\right)^2$, shows there are vertical asymptotes at $x = -C$ which are impossible to detect from the analysis in part i.

3. $y' = 1 - y$ has the equilibrium solution $y(x) = 1$. Nonequilibrium solutions are obtained from $\int \frac{1}{1-y}\,dy = \int dx$, which gives $-\ln|1-y| = x+C$, or $y(x) = 1 \pm e^{-C}e^{-x} = 1 + ce^{-x}$ where $c = \pm e^{-C} \neq 0$. Thus, the complete solution consists of two parts: $y(x) = 1$ and $y(x) = 1 + ce^{-x}$, where $c \neq 0$.

(a) $y(0) = 0$ does not lie on $y(x) = 1$ but does lie on $y(x) = 1 + ce^{-x}$, where $c \neq 0$. In fact, $0 = 1 + c$, so $c = -1$. Thus, $y(x) = 1 - e^{-x}$ is the solution.

(b) $y(0) = 1$ lies on $y(x) = 1$ so that $y(x) = 1$ is the solution.

(c) $y(0) = 2$ does not lie on $y(x) = 1$ but does lie on $y(x) = 1 + ce^{-x}$, where $c \neq 0$. In fact, $2 = 1 + c$, so $c = 1$. Thus, $y(x) = 1 + e^{-x}$ is the solution.

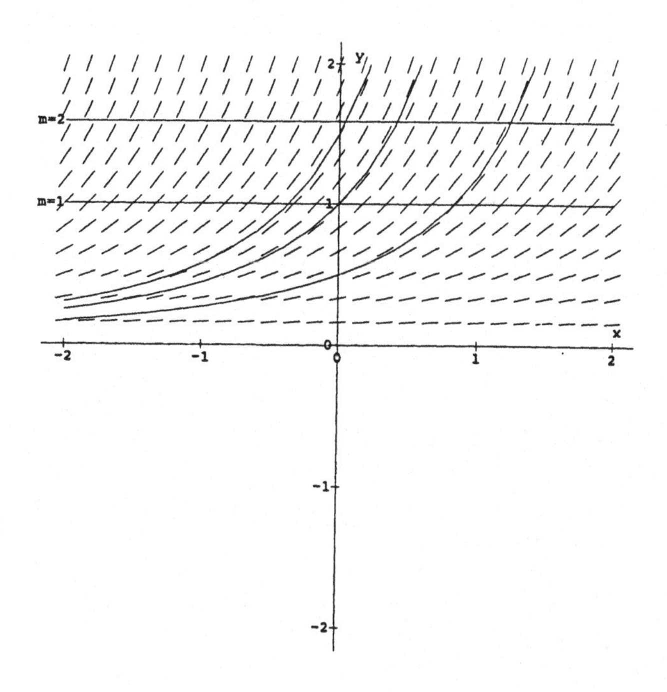

Figure 2-3 Exercise 1(e).

(d) If $y_0 = 1$ then $y(x_0) = y_0$ lies on $y(x) = 1$ so that $y(x) = 1$ is the solution.

If $y_0 \neq 1$ then $y(x_0) = y_0$ does not lie on $y(x) = 1$ but does lie on $y(x) = 1 + ce^{-x}$ where $c \neq 0$. In fact, $y_0 = 1 + ce^{-x_0}$, so $c = (y_0 - 1)\,e^{x_0}$. Thus, $y(x) = 1 + (y_0 - 1)\,e^{x_0 - x}$ is the solution.

5. **(a)** Figure 2.11 in the text is $y' = \frac{y^2-1}{y^2+1}$. For Figure 2.11 in the text, $m = 0$ at $y = \pm 1$, and $m = -1$ at $y = 0$. The graph is increasing when $y < -1$ or $y > 1$.

Figure 2.12 in the text is $y' = -\frac{y^2+1}{y^2-1}$. For Figure 2.12 in the text, the slopes are vertical at $y = \pm 1$, and $m = 1$ at $y = 0$. The graph is increasing when $-1 < y < 1$.

(b) You should notice that the two mystery slope fields are perpendicular to one another. In other words, their solution curves meet at 90° angles. This knowledge would have made the analysis in part (a) easier because the two differential equations must be negative reciprocals of one another in order for their solution curves to meet at 90° angles.

7. The slope field of the differential equation discussed in Chapter 1 — namely, $\frac{dy}{dx} = g(x)$ — cannot be symmetric about the x-axis. If y is replaced with $-y$, the differential equation changes to $-\frac{dy}{dx} = g(x)$. Therefore, the differential equation cannot be symmetric about the x-axis.

Example: $\frac{dy}{dx} = 4x$ Now, replace y with $-y$. $-\frac{dy}{dx} = 4x$.

9. $y' = -y^2$ has an equilibrium solution $y(x) = 0$. Nonequilibrium solutions are obtained from $\int \frac{1}{y^2}\,dy = -\int dx$, which gives $-y^{-1} = -x + C$, or $y(x) = 1/(x - C)$. Thus, the complete solution consists of two parts: $y(x) = 0$ and $y(x) = 1/(x - C)$.

Now consider the graph of $y(x) = 1/(x - C)$. If we start at a point which is a little above the x-axis, the solution curve will become closer to the x-axis as x increases.

However, if we start at a point which is a little below the x-axis, the solution curve will decrease, and become asymptotic to a vertical line.

Because in the semistable case solution curves on one side of an equilibrium solution move away from this equilibrium solution, it cannot be stable.

11. If we displace a pendulum which is hanging straight down from its equilibrium position, it will oscillate about that equilibrium position. If we displace a pendulum which is balancing vertically from its equilibrium position and release it, the pendulum will never return to that vertical position.

2.2 Simple Models

1. Because $y(t) = Ce^{kt}$ and $y(t_0) = y_0$ we have $y(x) = y_0 e^{k(t-t_0)}$. Let $t = t_0 + T_d$ be the time at which $y(t_0 + T_d) = 2y_0$, so that $2y_0 = y_0 e^{kT_d}$, that is, $e^{kT_d} = 2$. Thus, $T_d = \frac{1}{k}\ln 2$. The fact that T_d is independent of $y(t_0)$ tells us that no matter when we decide to start measuring, and no matter what the initial value of y, it will still take T_d for y to double.

3. Solving $(1/P)dP/dt = k$ gives $P(t) = Ce^{kt}$. If t measures time after 1800, then $P(0) = 1$, $P(100) = 1.7$. This gives $C = 1$, $k = \ln(1.7)/100 \approx 0.0053$, and $P(t) = \exp(0.0053t)$. Thus $P(200) = 2.89$ billion gives the prediction of the world's population in the year 2000. The doubling time is $\ln 2/k \approx 131$ years. As the world's population in 1990 was 5.3 billion with a doubling time of about 40 years, this is not a great prediction.

5. From $dC/dt = kC$, we solve for C and obtain $C(t) = C_0 e^{kt}$. In 10 hours the concentration is $C_0/2$, so we find k from $1/2 = e^{10k}$ as $k = -(\ln 2)/10 \approx -0.069$. The concentration of the drug will be 10% of its original value when $C(t) = 0.1C_0$. Thus, we solve $0.1C_0 = C_0 e^{kt}$ for t, and we find $t = \ln 0.1/k = -10\ln 0.1/\ln 2 \approx 33.2$ hours.

7. We use a model of Houston's population as $P(t) = Ce^{kt}$, where t is number of years after 1850.

   If we require that this function match the data in the years 1850 and 1970, we find that $C = 18.632$ and $k = \ln(1999.316/18.632)/140 \approx 0.039$. Plotting $P(t) = 18.632\exp(0.039t)$ along with the data, or evaluating $P(t)$ for several times between $t = 10$ and $t = 130$ shows good agreement.

   Alternatively, we plot $\ln P$ versus t obtaining an approximate straight line with slope 0.0379 and intercept 3.0437, so we have $P(t) = \exp(0.0379t + 3.0437) = 20.98\exp(0.0379t)$. Plotting $P(t) = 20.98\exp(0.0379t)$ along with the data, or evaluating $P(t)$ for several times between $t = 10$ and $t = 130$ also shows good agreement.

   In the next calculation we will use $P(t) = 18.632\exp(0.039t)$.

   A model of the world's population with 5.3 billion in 1990 and a doubling time of 40 years is $W(t) = 5.3e^{kt}$, where $k = \ln 2/40 \approx 0.0173$. For comparison sake, we rewrite Houston's population (in billions) as $P(t) = 0.000018632\exp[0.039(t + 140)]$, where t is the number of years after 1990. The value of t where these two populations are equal may be found by solving $W(t) = P(t)$ for t. Using logarithms we find that $\ln 5.3 + 0.0173t = \ln 0.000018632 + 0.039(t + 140)$, giving $t \approx 327$ years. (We may also solve these equations by graphing $W(t)$ and $P(t)$ and seeing where the two graphs intersect.) If we use $P(t) = 20.98\exp(0.0379t)$, we find $t \approx 346$ years.

9. Let $x(t)$ be the distance in meters that the bullet has travelled from time $t = 0$ when the bullet enters the wood. Thus, $x(0) = 0$ and $v(t) = dx/dt$ has the property that $v(0) = 200$.

If we let T be the time when the bullet emerges from the wood, we have $x(T) = 0.1$ and $v(T) = 80$.

From $dv/dt = -kv^2$ we find $v(t) = 1/(kt + C)$, which, because $v(0) = 200$, gives $C = 1/200$, and so

$v(t) = 200/(200kt + 1)$.

This gives $dx/dt = 200/(200kt + 1)$, which can be integrated to yield

$x(t) = (1/k)\ln(200kt + 1) + C$.

From $x(0) = 0$ we find $C = 0$, so that $x(t) = (1/k)\ln(200kt + 1)$.

Because $v(T) = 80$ we find $80 = 200/(200kT + 1)$, or $200kT + 1 = 200/80 = 5/2$.

Because $x(T) = 0.1$ gives $0.1 = (1/k)\ln(200kT + 1)$, we find $k = 10\ln(200kT + 1) = 10\ln(5/2)$, so $T = 1.5/(200k) = 1.5/[2000\ln(5/2)] \approx 8.19 \times 10^{-4}$ sec.

The bullet can never come to rest in finite time, because $v(t) = 200/(200kt + 1) \neq 0$. Thus, according to this model, no thickness of wood can bring the bullet to rest. The model is not very realistic.

2.3 The Logistic Equation

1. Because the "carrying capacity" of the rumor is 35,000 then $b = 35000$.

The solution of $y' = ay(b - y)$ is $y(t) = by_0/\left[(b - y_0)e^{-abt} + y_0\right]$.

According to the information given, $y_0 = 35$ and $y(1) = 700$, so we can calculate a from $700 = 35000 \cdot 35/\left(34965e^{-a35000} + 35\right)$. This gives $a \approx 8.61 \times 10^{-5}$.

We now must find the time T when $y(T) = 0.9 \times 35000 = 31500$, which will be when $31500 = 35000 \cdot 35/\left(34965e^{-8.61\times10^{-5}\times35000T} + 35\right)$. Solving for T gives $T \approx 3.02$ days.

3. If $y(x) = 1/u(x)$, then $\frac{dy}{dx} = -\frac{1}{u^2}\frac{du}{dx}$, so $y' = ay(b - y)$ becomes $-\frac{1}{u^2}\frac{du}{dx} = \frac{a}{u}(b - \frac{1}{u})$. Multiplying both sides of this equation by $-u^2$ gives $\frac{du}{dx} = a(1 - bu)$. Note that $u = 1/b$ is an equilibrium solution, which corresponds to $y(x) = b$. If $u \neq 1/b$, we may rewrite this equation as $1/(1 - bu)\frac{du}{dx} = a$ and integrate both sides with respect to x to obtain $-(1/b)\ln|1 - bu| = ax + c$. Multiplying by $-b$ and exponentiating both sides gives $u = [1 - C\exp(-abx)]/b$. Taking reciprocals of both sides gives an answer which is equivalent to that given in (2.26). These details are easier than those used to derive (2.26).

5. The curve labelled A has its inflection below the horizontal line equal to 1/2 of the equilibrium solution, so it cannot be a solution of a logistic equation.

7. We first note that the two equilibrium solutions are $y = 1$ and $y = -1$. If $1 - y^2 \neq 0$, we write $\frac{dy}{dx} = (1/2)(1 - y^2)$ as $2/(1 - y^2)\frac{dy}{dx} = 1$. We then use partial fractions to change the form of the left-hand side and find $[1/(1 - y) + 1/(1 + y)]\frac{dy}{dx} = 1$. If we integrate both sides with respect to x we obtain $-\ln|1 - y| + \ln|1 + y| = x + c$. We may write this as $(1 + y)/(1 - y) = Ce^x$ or, solving for y, $y(x) = (Ce^x - 1)/(Ce^x + 1)$.

(a) If $y(0) = 0$, as $y(x) = 0$ is not an equilibrium solution, we substitute $(0,0)$ into this last equation to obtain $C = 1$, and our solution to this initial value problem is $y(x) = (e^x - 1)/(e^x + 1)$.

(b) Because $y(0) = 1$ lies on the equilibrium solution $y(x) = 1$, $y(x) = 1$ is the solution of this initial value problem.

(c) Because $y(0) = -1$ lies on the equilibrium solution $y(x) = -1$, $y(x) = -1$ is the solution of this initial value problem.

9. **(a)** This is the logistic equation with $a = \beta$ and $b = \alpha/\beta$. The equilibrium solutions are $y(x) = 0$, $y(x) = \frac{\alpha}{\beta}$. Nonequilibrium solution

$$y(x) = \frac{\alpha y_0}{(\alpha - \beta y_0)\, e^{-\alpha x} + \beta y_0}.$$

(b) The differential equation $y' = \alpha y$ has solutions $y(x) = 0$ and $y(x) = y_0 e^{\alpha x}$, and the second is contained in part (a) with $\beta = 0$.

(c) The differential equation $y' = -\beta y^2$ has solutions $y(x) = 0$ and

$$y(x) = \frac{y_0}{\beta y_0 x + 1},$$

and the second is not contained in part (a) with $\alpha = 0$.

(d) The equilibrium solutions are $y(x) = 0$, $y(x) = \frac{\alpha}{\beta}$. Only the equilibrium solution $y(x) = 0$ is included in the results of parts (b) and (c).

11. **(a)** The slope field for this equation is reminiscent of a logistic equation for $y > 0$.

(b) $y' = -ky \ln \frac{y}{b}$, $y'' = -ky' \ln \frac{y}{b} - ky' = -k\left(\ln \frac{y}{b} + 1\right) y'$. Recall that $y \geq 0$ on physical grounds — it is the size of the tumor.

The function will be increasing ($y' > 0$) when $\ln \frac{y}{b} < 0$ — that is, when $y < b$ — and will be decreasing when $y > b$.

The function will be concave up ($y'' > 0$) when $\left(\ln \frac{y}{b} + 1\right) y' < 0$. The term $\ln \frac{y}{b} + 1$ is zero when $y = be^{-1}$ and so $\ln \frac{y}{b} + 1 > 0$ when $y > be^{-1}$ and $\ln \frac{y}{b} + 1 < 0$ when $y < be^{-1}$. Combining these facts we find that $y(t)$ is concave up when $y < be^{-1}$ or $y > b$. It is concave down when $be^{-1} < y < b$. Points of inflection occur when $y = be^{-1}$.

(c) For $y > 0$, the equilibrium solution is $y(t) = b$.

(d) For nonequilibrium solutions $\int \frac{1}{y \ln \frac{y}{b}}\, dy = -k \int dt$. In order to integrate the left-hand side we make the substitution $u = \ln \frac{y}{b}$, in which case we have $\int \frac{1}{y \ln \frac{y}{b}}\, dy = \int \frac{1}{u}\, du$.

Integration thus yields $\ln |u| = -kt + C$, or $\ln \left|\ln \frac{y}{b}\right| = -kt + C$.

Solving for $y(t)$ gives $y(t) = b \exp\left(ce^{-kt}\right) = b\,(\exp c)^{e^{-kt}}$, where $c = \pm e^C \neq 0$.

Because $y(0) = b \exp c$, we thus have $y(t) = b\left(\frac{y(0)}{b}\right)^{e^{-kt}}$.

(e) The inflection point occurs when $be^{-1} = y(t) = b\left(\frac{y(0)}{b}\right)^{e^{-kt}}$. Solving this for t gives $t = \frac{1}{k} \ln\left[-\ln\left(\frac{y(0)}{b}\right)\right]$. For the right-hand side to be defined we need $-\ln\left(\frac{y(0)}{b}\right) > 0$, that is, $y(0) < b$. For $t > 0$ we further need $-\ln\left(\frac{y(0)}{b}\right) > 1$, that is, $y(0) < be^{-1}$. Thus, there will be no inflection point if $y(0) > be^{-1}$.

(f) The same initial conditions and the same carrying capacity mean that $y(0)$ and b are the same in the two equations.

The point of inflection in the logistic equation occurs at $y = b/2$, so $b/2 = by_0/\left[(b-y_0)e^{-abt} + y_0\right]$, which, when solved for t, gives $t = \ln\left(\frac{b-y_0}{y_0}\right)/(ab)$. The point of inflection in the Gompertz equation occurs when $t = \frac{1}{k}\ln\left[-\ln\left(\frac{y(0)}{b}\right)\right]$. If these are to be the same then $\ln\left(\frac{b-y_0}{y_0}\right)/(ab) = \frac{1}{k}\ln\left[-\ln\left(\frac{y(0)}{b}\right)\right]$, so $k = ab\ln\left[-\ln\left(\frac{y(0)}{b}\right)\right]/\ln\left(\frac{b-y_0}{y_0}\right)$.

The graphs are very similar except that inflection point of the Gompertz equation occurs at $be^{-1}\,(= b/e)$ while that of the logistic equation occurs at $b/2$, which is above b/e.

13. The logistic equation is $dy/dx = ay(b-y)$ with solution (from (2.27)) $y(x) = by_0/[(b-y_0)e^{-abx} + y_0]$. The differential equation we used for Botswana was $dy/dx = ky$, with solution $y(x) = Ce^{kx}$ with $C = 0.755$ and $k = 0.0355$. Comparing the two we need $ab = 0.0355$ and $y_0 = 0.755$, so our logistic solution becomes $y(x) = 0.755/[(1 - 0.755/b)e^{-0.0355x} + 0.755/b]$. If we graph this function for several values of b such that $0.755/b$ is small when compared with 1 (like $b = 20$ or 30) we see very little difference from $y(x) = 0.755e^{-0.0355x}$ for $0 < x < 15$.

15. **(a)** From $y(x) = 12.94/\left(e^{-0.087x} + 0.0498\right)$ we have $\lim_{x\to\infty} y(x) = 12.94/0.0498$. Thus the rise time is given by the value of T such that $y(T) = 12.94/\left(e^{-0.087T} + 0.0498\right) = 0.99 \times 12.94/0.0498$. Using logarithms we find $T = -1/0.087\ln(0.0498/990) \approx 87.3$ days.

(c) Using $b = 10000$, $a = 0.0001$, and $y_0 = 100$ in (2.27) gives $y(x) = 10^6/\left(9900e^{-x} + 100\right)$. Here $\lim_{x\to\infty} y(x) = 10^4$, so we need to find T such that $y(T) = 10^6/\left(9900e^{-T} + 100\right) = 0.99 \times 10^4$. Solving for T gives $T = 2\ln 99 \approx 9.2$ years.

17. We use the solution for the logistic equation as $y(x) = bC/[\exp(-abx) + C]$. If we estimate the limiting value of deaths as 9100, that is the value of b. We then follow the example that gave (2.29) for sunflower growth and graph $\ln[y/(9100-y)]$ against x, using the data in Table 2.10. The slope of this line is ab, so $ab = 0.3968$ and the y-intercept is $c = -7.0612$, so $C = \exp(-7.0612)$. Plotting this function on the same graph as the data (or evaluating this function in specific years) shows this is a very good model. See Figure 2-4.

19. Because $X = abx$ we have, by the chain rule, $\frac{dy}{dx} = \frac{dy}{dX}\frac{dX}{dx} = \frac{dy}{dX}ab$ — that is, $\frac{dy}{dx} = ab\frac{dy}{dX}$. Because $y = bY$ this becomes $\frac{dy}{dx} = ab\frac{d}{dX}(bY) = ab^2\frac{dY}{dX}$. We substitute this identity and $y = bY$ into $y' = ay(b-y)$ to find $ab^2\frac{dY}{dX} = abY(b - bY)$, that is, $\frac{dY}{dX} = Y(1-Y)$.

To solve $\frac{dY}{dX} = Y(1-Y)$ for $Y(X)$ we first write down the equilibrium solutions, namely $Y(X) = 0$ and $Y(X) = 1$. For the nonequilibrium solutions we have $\int \frac{1}{Y(1-Y)}\,dY = \int dX$, so $\int\left(\frac{1}{Y} + \frac{1}{1-Y}\right)dY = X + C$. This yields $\ln\left|\frac{Y}{1-Y}\right| = X + C$, or $Y(X) = \frac{ce^X}{1+ce^X} = \frac{c}{e^{-X}+c}$, where $c = \pm e^C \neq 0$.

If we substitute $X = abx$ and $Y = \frac{y}{b}$ into the equilibrium solutions we have $y(x) = 0$ and $y(x) = b$, and in the nonequilibrium case we find $y(x) = \frac{bc}{e^{-abx}+c}$, which agrees with (2.29).

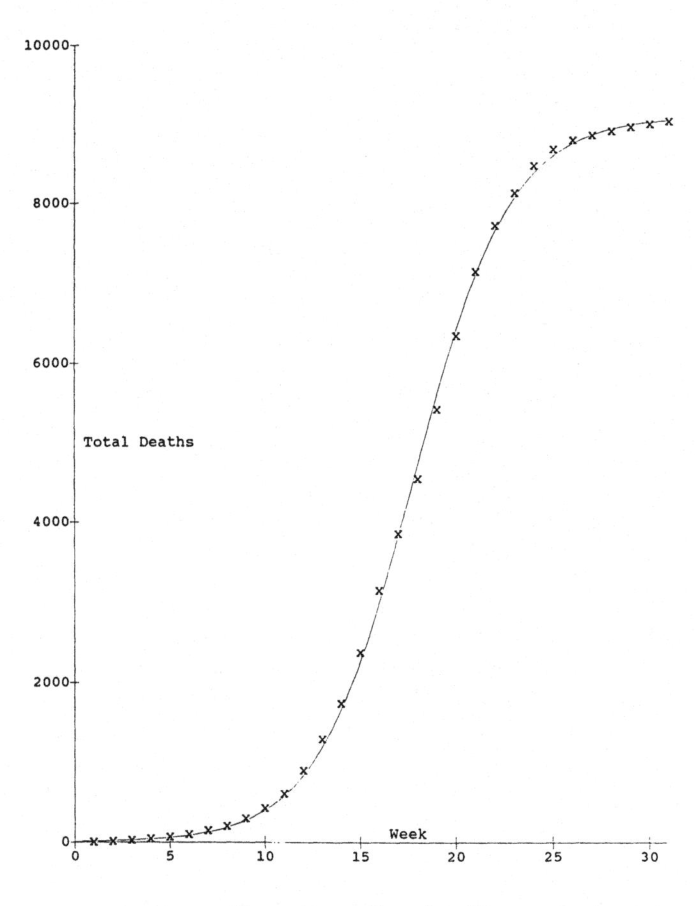

Figure 2-4 Exercise 17.

2.4 Existence and Uniqueness of Solutions, and Words of Caution

1. The constant function $y(x) = \frac{2}{3}$ satisfies the differential equation and is the only equilibrium solution. The initial condition $y(x_0) = \frac{2}{3}$ for any x_0 will pick out the solution $y(x) = \frac{2}{3}$.

3. The person is right.

The equilibrium solution is $y(x) = 0$ while the nonequilibrium solutions are obtained from $\int y^{-1/3}\, dy = x + C$, namely, $y(x) = \pm\left[\frac{2}{3}(x + C)\right]^{3/2}$, which is defined for $x + C \geq 0$.

We can make $y(x) = \pm\left[\frac{2}{3}(x + C)\right]^{3/2}$ pass through $(0, 0)$ by setting $C = 0$, and we find $y(x) = \pm\left(\frac{2}{3}x\right)^{3/2}$ for $x > 0$.

There are three "obvious" solutions through $(0, 0)$, namely, $y(x) = 0$ for all x; $y(x) = 0$ for $x \leq 0$, followed by $y(x) = \left(\frac{2}{3}x\right)^{3/2}$ for $x > 0$; and $y(x) = 0$ for $x \leq 0$, followed by $y(x) = -\left(\frac{2}{3}x\right)^{3/2}$ for $x > 0$.

In fact, there are more solutions. For example $y(x) = 0$ for $x \leq 1$, followed by $y(x) = \left[\frac{2}{3}(x - 1)\right]^{3/2}$ for $x > 1$.

5. Because $ay(b - y)$ satisfies the conditions of the Existence-Uniqueness theorem for all values of x and y, all solutions are unique. Therefore no other solution may intersect the equilibrium solution. Other solutions may approach the equilibrium solution as $x \to \infty$, but not touch it.

7. **(a)** If c is very large, many animals are being poached and the population may die out.

(b) If $y = bY/2$, and $x = 2X/(ab)$,
$dy/dx = d(bY/2)/dx = (b/2)dY/dx = (b/2)(dY/dX)(dX/dx) = (b/2)(dY/dX)(ab)/2$.
This means $dy/dx = ay(b-y) - c$ becomes
$(ab^2/4)(dY/dX) = (abY/2)[b - bY/2] - c$, or
$dY/dX = Y(2-Y) - C$, where $C = 4c/(ab^2)$.

(c) $dY/dX = Y(2-Y) - C$ has equilibrium solutions where $Y^2 - 2Y + C = 0$, or
$Y = \left(2 \pm \sqrt{4-4C}\right)/2 = 1 \pm \sqrt{1-C}$.
i. If $C > 1$ there are no equilibrium solutions. ii. If $C = 1$ there is one (repeated) equilibrium solution. iii. If $C < 1$ there are two equilibrium solutions: one larger than 1, one smaller than 1.

(d) If $C = 1$, the differential equation is $dY/dX = -(1-Y)^2$, so solutions are always decreasing and the equilibrium solution is semistable. If $C < 1$, we may write the differential equation as $dY/dX = -(Y - r_1)(Y - r_2)$, where $r_1 = 1 - \sqrt{1-C}$ and $r_2 = 1 + \sqrt{1-C}$. If $Y > r_2$, $dY/dX < 0$, while if $r_1 < Y < r_2$ we have $dY/dX > 0$. Thus $Y = r_2$ is a stable equilibrium solution. If $Y < r_1$, $dY/dX < 0$, so $Y = r_1$ is an unstable equilibrium solution.

(e) If $C > 1$, we may write the differential equation as $dY/dX = -(1-Y)^2 - C + 1$, giving Y as a decreasing function of X. As there are no equilibrium solutions, Y will always decrease, eventually reaching a value of 0.

(f) If $C = 1$, the population will die out if the initial condition gives $Y(0) < 1$. If $C < 1$, the population will die out if the initial condition gives $Y(0) < r_1$. (Notice it does not make sense to have $Y(0) < 0$.) The population will not die out, but have a limiting value of $1 + \sqrt{1-C}$ if $Y(0) > r_1 = 1 - \sqrt{1-C} > 0$.

9. (a) No. If the equilibrium solutions $y(x) = 1$, $y(x) = 2$, and $y(x) = 3$ were all stable, then $y(x)$ would have to be an increasing function for $y < 2$ and a decreasing function for $y > 1$. Because solutions of autonomous differential equations have slopes which do not change sign between equilibrium solutions, this is impossible.

(b) No. The same reasoning for part (a) applies here.

(c) Yes. One example is $y' = C(1-y)(2-y)^2(3-y)^2$, where $C > 0$. Many other examples are possible.

11. The nonequilibrium solution of $P' = rP^n$ is $P(t) = 1/\left[C - (n-1)rt\right]^{1/(n-1)}$. With $P(0) = P_0$ we find $C = P_0^{n-1} > 0$. Because $n > 1$, $P(t)$ has a vertical asymptote at $C - (n-1)rt = 0$, that is, $t = C/\left[r(n-1)\right]$. Because $C > 0$, this time occurs after $t = 0$. Thus, $P(t) \to \infty$ as $t \to C/\left[r(n-1)\right]$, so that the new investor's account grows to infinity in finite time. Not good for the old investors!

2.5 Qualitative Behavior of Solutions Using Phase Lines

1. The equilibrium solution of $y' = -ky\ln(y/b)$ is $y(x) = b$. If $y > b$, $y' < 0$ and if $0 < y < b$, $y' > 0$, so $y(x) = b$ is a stable equilibrium solution. The differential equation is only defined for $y > 0$, and because $y' > 0$ for $0 < y < b$, $y(x) = 0$ cannot be a stable equilibrium. The logistics equation has these same characteristics regarding it's equilibrium solutions.

3. The equilibrium solutions are $P(t) = 0$, $P(t) = b$, and $P(t) = c$, where $b < c$. The solutions 0 and c are stable, while b is unstable. Thus, if initially $P(0) > b$ then

$P(t) \to c$, the carrying capacity for passenger pigeons. However, if $0 < P(0) < b$ then $P(t) \to 0$ and the pigeons become extinct. It is a reasonable model. The quantity b is the threshold for survival.

5. **(a)** Yes, because $(y-a)(y-b)(y-c)$ and its derivative with respect to y are continuous.

(b) If we substitute $y(x) = a$ into the equation $y' = (y-a)(y-b)(y-c)$ we find it is satisfied. Similar arguments apply to $y(x) = b$ and $y(x) = c$.

(c) i If $a = b = c = 1$ the only equilibrium solution is $y(x) = 1$. Other solutions will move away from $y(x) = 1$ so it is unstable.

ii If $a = b = 1$ and $c = -1$ there are two equilibrium solutions, $y(x) = 1$ and $y(x) = -1$. Other solutions near $y = -1$ move away from $y(x) = -1$ so $y(x) = -1$ is unstable. Solutions near $y(x) = 1$ approach it if $y < 1$ but leave it if $y > 1$. Thus, $y(x) = 1$ is semistable.

iii Similarly, in the case $a = 1$, $b = 0$, $c = -1$, $y(x) = -1$ and $y(x) = 1$ are unstable, but $y(x) = 0$ is stable.

7. **(a)** If $a < 0$, y' is always negative, so there are no equilibrium solutions. If $a = 0$, y' is never positive, so the equilibrium solution, $y(x) = 0$ is semistable. If $a \geq 0$, $y' = (\sqrt{a} - y)(\sqrt{a} + y)$ so there are two equilibrium solutions, $y(x) = \pm\sqrt{a}$. Computing $dg(y)/dy$ gives $dg/dy = -2y$. Using The Derivative Test for Stable or Unstable Equilibrium Solutions shows that $y(x) = \sqrt{a}$ is stable while $y(x) = -\sqrt{a}$ is unstable. Thus dramatic changes occur when $a = 0$.

(b) $y' = y(a - y^2)$ has three equilibrium solutions $(0, \sqrt{a}, -\sqrt{a})$, if $a > 0$ and one $y(x) = 0$, if $a \leq 0$. Thus the number of equilibrium points changes when $a = 0$. Computing $dg(y)/dy$ gives $dg/dy = a - 3y^2$. Using The Derivative Test for Stable or Unstable Equilibrium Solutions shows that $y(x) = 0$ is stable if $a < 0$ and unstable $a > 0$. If $a = 0$, the equilibrium solution $y(x) = 0$ is stable because our solutions are always decreasing for $y > 0$ and increasing for $y < 0$. Using the same theorem shows that both $y(x) = \sqrt{a}$ and $y(x) = -\sqrt{a}$ are stable.

9. If we sketch $g(y)$ vs y, and $g(y)$ has two equilibrium points, then $g(y)$ is equal to zero at these two points. Between these two equilibrium points, $g(y)$ must always be either positive or negative. Because $g(y)$ has a continuous derivative, $g(y)$ will have a maximum value or a minimum value between these two points. At these maximum or minimum values the slope of our solution curves changes sign (recall $y' = g(y)$) giving an inflection point.

11. A possible differential equation is $y' = a(b-y)(c-y)$, where a, b, and c are positive constants. Possible values of these constants are $a = 0.03$, $b = 276$, and $c = 211$, which give the ultimate world record as 211 seconds.

13. **(a)** The equilibrium solutions are $y(x) = 0$ and $y(x) = 4$. They are both stable.

(b) The nonequilbrium solutions are obtained from

$$\int \frac{4-2y}{y^2-4y}\,dy = \int 1\,dx,$$

namely, $-\ln\left|y^2 - 4y\right| = x + C$, or
$y^2 - 4y = Ae^{-x}$, where $A = \pm e^{-C}$.
With $y(0) = y_0$, we have $y_0^2 - 4y_0 = A$, so
$y^2 - 4y = \left(y_0^2 - 4y_0\right)e^{-x}$. Solving for $y(x)$ gives
$y(x) = 2 \pm \sqrt{4 + \left(y_0^2 - 4y_0\right)e^{-x}}$.
Note that if we put $x = 0$ in this equation then

$y_0 = 2 \pm \sqrt{4 + (y_0^2 - 4y_0)} = 2 \pm \sqrt{(y_0 - 2)^2} = 2 \pm |y_0 - 2|$. For this to be an identity in y_0, we see that the + sign applies when $y_0 > 2$, and the − sign applies when $y_0 < 2$. Thus,

$y(x) = 2 + \sqrt{4 + (y_0^2 - 4y_0)\, e^{-x}}$ if $y_0 > 2$.

$y(x) = 2 - \sqrt{4 + (y_0^2 - 4y_0)\, e^{-x}}$ if $y_0 < 2$.

As $x \to \infty$, $y(x) \to 4$ if $y_0 > 2$ and $y(x) \to 0$ if $y_0 < 2$. This is consistent with part (a).

(c) When we try to draw the phase line, arrows point towards 0 and towards 4, because they are both stable. Thus between 0 and 4, we have arrows pointing in both directions.

(d) Plotting $(y^2 - 4y) / (4 - 2y)$ versus y shows we have a vertical asymptote at $y = 2$, which means that $(y^2 - 4y) / (4 - 2y)$ is not continuous between 0 and 4 so our usual phase line analysis fails.

2.6 Bifurcation Diagrams

1. **(a)** The equilibrium solutions occur when $a = y^2$, that is, $y = \pm\sqrt{a}$, for $a \geq 0$. Also $\frac{d}{dy}(a - y^2) = -2y$, which is positive (meaning unstable) if $y = -\sqrt{a}$, and negative (meaning stable) if $y = \sqrt{a}$. See Figure 2-5.

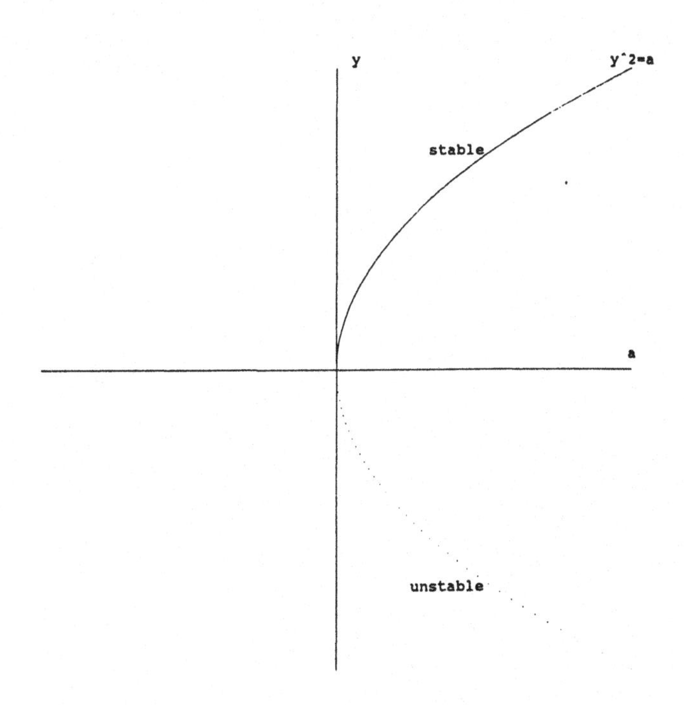

Figure 2-5 Exercise 1(a).

(c) The equilibrium solutions occur when $y^2 + y + a = 0$, that is, $y = -\frac{1}{2} \pm \sqrt{\frac{1}{4} - a}$, for $a \leq \frac{1}{4}$. Also $\frac{d}{dy}(y^2 + y + a) = 2y + 1$, which is positive (meaning unstable) if $y = -\frac{1}{2} + \sqrt{\frac{1}{4} - a}$, and negative (meaning stable) if $y = -\frac{1}{2} - \sqrt{\frac{1}{4} - a}$. See Figure 2-6.

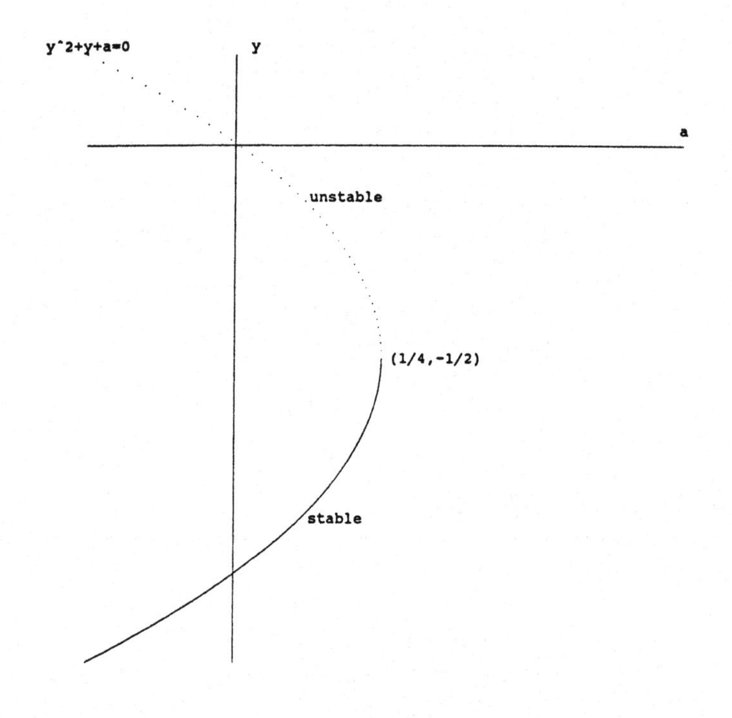

Figure 2-6 Exercise 1(c).

3. FIRST ORDER DIFFERENTIAL EQUATIONS — QUALITATIVE AND QUANTITATIVE ASPECTS

3.1 Graphical Solutions Using Calculus

1. **(a)** If $u(x) = y(x) - x$, $u' = y' - 1$. Using these values in $y' = x - y$ gives $u' + 1 = -u$.

(b) The only equilibrium solution we have is $u = -1$. Writing $u' + 1 = -u$ as $[1/(1+u)]du/dx = -1$ and integrating gives $\ln|1+u| = -x + c$. We may write this as $1 + u = Ce^{-x}$, where $C = \pm e^c$, or as $u(x) = -1 + Ce^{-x}$. Allowing C to be 0 includes the equilibrium solution.

(c) Substituting $u(x) = -1 + Ce^{-x}$ into $u(x) = y(x) - x$ gives $y(x) = x - 1 + Ce^{-x}$.

3. **(a)** $y' = nx^{n-1}$, so $xy' - 3y = 0$ becomes
$nx^n - 3x^n = (n-3)x^n = 0$.
$n = 3$.

(c) $y' = n\exp(-nx)$ so $y' + 32y - 32 = 0$ becomes
$n\exp(-nx) + 32[1 - \exp(-nx)] - 32 = (n - 32)\exp(-nx) = 0$, so
$n = 32$.

(e) $y' = C\exp(1 - n\exp(x))(-n\exp(x))$, so $y' + \exp(x)y = 0$ becomes
$C\exp(1 - \exp(x))[-n\exp(x) + \exp(x)] = 0$, so
$n = 1$.

5. **(a)** If $y(x) = Cx^4 - 1$ is to pass through $(0, -1)$ then $-1 = C \times 0 - 1$, so C may be any real number.

(c) If $y(x) = C(x-2)^2e^x$ is to pass through $(0, -1)$ then $-1 = C \times 4$, so $C = -1/4$.

7. **(a)** $y' = \frac{x}{y}$ is symmetric across the x- and y-axes, and about origin.

(b) Isoclines: $\frac{x}{y} = m$. If $m = 0$ the isocline is $x = 0$, otherwise they are $y = \frac{x}{m}$.

(c) The isoclines $y = \frac{1}{m}x$ are straight lines through the origin with slope $1/m$.

(d) y is increasing for $x/y > 0$ — that is, in the first and third quadrants — and y is decreasing for $x/y < 0$ — that is, in the second and fourth quadrants.
$y'' = \frac{1}{y^2}(y - xy') = \frac{1}{y^2}\left(y - x\frac{x}{y}\right) = \frac{1}{y^3}(y^2 - x^2) = \frac{1}{y^3}(|y| + |x|)(|y| - |x|)$.
Concave up when $y > 0$ and $|y| > |x|$, or $y < 0$ and $|y| < |x|$.
Concave down when $y < 0$ and $|y| > |x|$, or $y > 0$ and $|y| < |x|$. See Figure 3-1.

(f) The point $x = 0$, $y = 1$ satisfies the equation $y^2 - x^2 = 1$. Also, by differentiating $y^2 - x^2 = 1$ with respect to x we have $2yy' - 2x = 0$, that is, $yy' = x$.

9. **(a)** $y' = -2xy$ is symmetric across the x- and y-axes, and about origin. Isoclines: $-2xy = m$. If $m = 0$ the isocline is $x = 0$ and $y = 0$, otherwise they are $y = -m/(2x)$. The isoclines $y = -m/(2x)$ are hyperbolas.

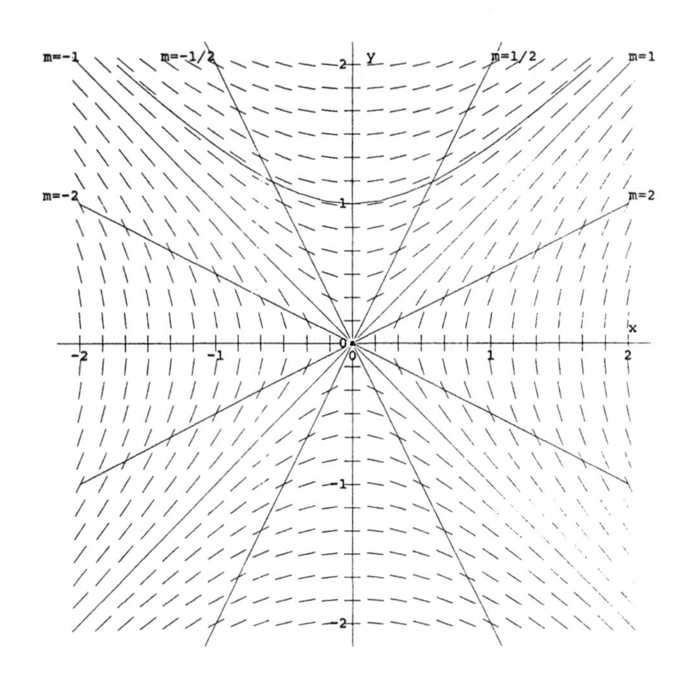

Figure 3-1 Exercise 7.

y is increasing for $-2xy > 0$ — that is, in the second and fourth quadrants — and y is decreasing for $-2xy < 0$ — that is, in the first and third quadrants.
$y'' = -2\,(y + xy') = -2\,[y + x\,(-2xy)] = -2y\,(1 - 2x^2)$.
Concave up when $y > 0$ and $|x| > 1/\sqrt{2}$, or $y < 0$ and $|x| < 1/\sqrt{2}$.
Concave down when $y < 0$ and $|x| > 1/\sqrt{2}$, or $y > 0$ and $|x| < 1/\sqrt{2}$. See Figure 3-2.

11. **(a)** It appears that solutions have a relative maximum along the isocline for zero slope, namely the curve $y = 0.5e^{-x}$. See Figure 3-3.

(b) At horizontal asymptotes, $\lim_{x\to\infty} y(x) =$ constant and $\lim_{x\to\infty} y'(x) = 0$. From the differential equation we see that $\lim_{x\to\infty} y'(x) = \lim_{x\to\infty} e^{-x} - \lim_{x\to\infty} 2y(x) = 0$, so the only possible horizontal asymptote is $y = 0$.
Also, $y'' = -e^{-x} - 2y' = -3e^{-x} + 4y$.
Thus, as $x \to \infty$,
if $y > 0.75e^{-x}$, solutions are concave up and decreasing,
if $0.5e^{-x} < y < 0.75e^{-x}$, solutions are concave down and decreasing,
if $y < 0.5e^{-x}$, solutions are concave down and increasing.
All this is consistent with having $y = 0$ as a horizontal asymptote.
The derivative is defined for all finite values of x and y, but that does not rule out having a vertical asymptote.

(c) From $y(x) = e^{-x} + (y_0 - 1)\,e^{-2x}$ we find
$y'(x) = -e^{-x} - 2\,(y_0 - 1)\,e^{-2x}$, so
$y' + 2y - e^{-x} = -e^{-x} - 2\,(y_0 - 1)\,e^{-2x} + 2\left[e^{-x} + (y_0 - 1)\,e^{-2x}\right] - e^{-x} = 0$ and
$y(0) = e^{-0} + (y_0 - 1)\,e^{-2\times 0} = y_0$.
Notice that for every finite value of y_0, $\lim_{x\to\infty} y(x) = 0$, confirming that $y = 0$ is a horizontal asymptote. From this form of $y(x)$ we see there are no vertical asymptotes.

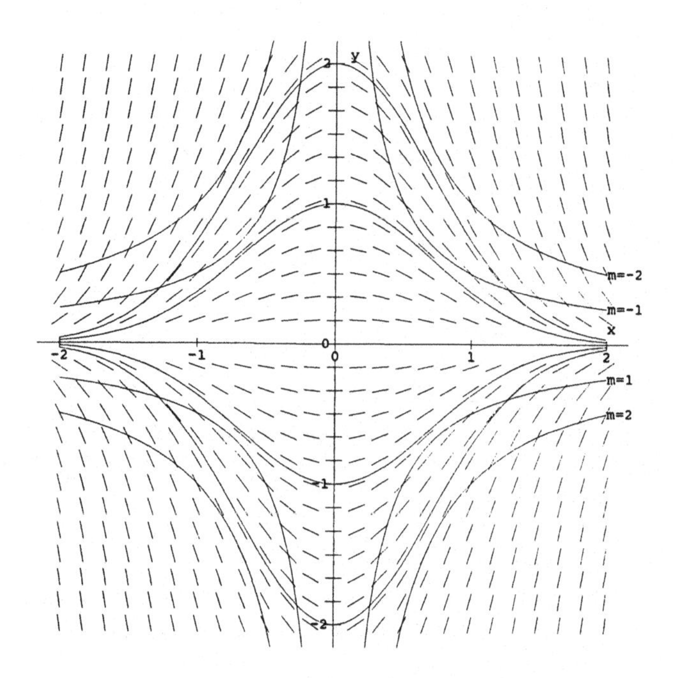

Figure 3-2 Exercise 9(a).

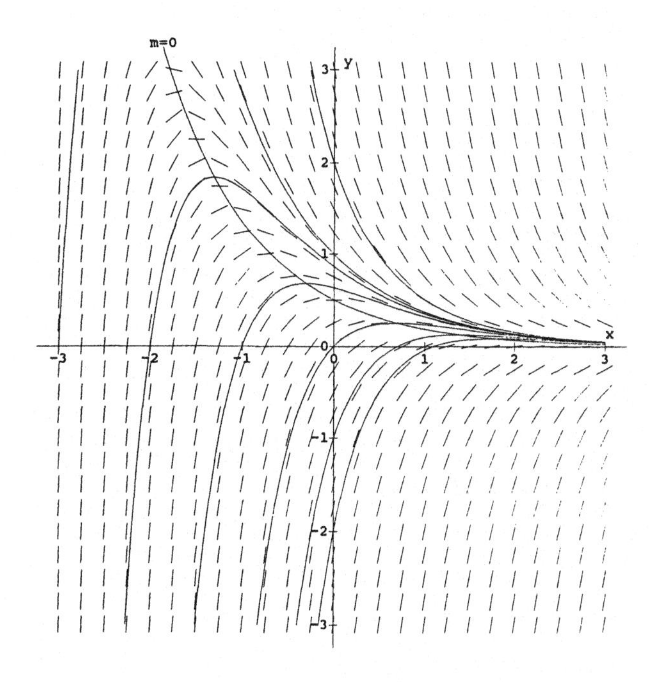

Figure 3-3 Exercise 11.

13. (a) In Figure 3.16 of the text an isocline for a slope of -1 appears to be $y = -x$. Thus our only two choices are $y' = -(x^2+1)/(y^2+1)$ or $y' = -(y^2+1)/(x^2+1)$. Because for $y = 1$ and x positive the slopes increase as x increases, the equation must be $y' = -(y^2+1)/(x^2+1)$.

(b) In Figure 3.17 of the text an isocline for a slope of 1 appears to be $y = x$. Thus our only two choices are $y' = (x^2+1)/(y^2+1)$ or $y' = (y^2+1)/(x^2+1)$. Because for $y = 0$ and x positive the slopes increase as x increases, the equation must be $y' = (x^2+1)/(y^2+1)$.

15. If we differentiate $x^2+y^2 = C$ with respect to x, we obtain $2x+2yy' = 0$, or $y' = -x/y$.

17. Differentiating $(x^2+y^2)^2 = 12(x^2-y^2)$ gives

$$2(x^2+y^2)(2x+2yy') = 24x - 24yy'$$

$$4x^3 + 4x^2yy' + 4xy^2 + 4y^3y' = 24x - 24yy'$$

$$4x^3 + 4xy^2 - 24x = (-24y - 4x^2y - 4y^3)y'$$

$$\frac{4x(x^2+y^2-6)}{-4y(6+x^2+y^2)} = y'$$

$$y' = \frac{x(6-x^2-y^2)}{y(6+x^2+y^2)}$$

Near the origin, the values of x and y are close to 0 so $6 \gg x^2+y^2$ and we can ignore x^2+y^2 in comparison with 6. Therefore, $y' \approx \frac{x}{y}$ (refer to Exercise 14) and the curves look like hyperbolas.

Away from the origin where $x^2+y^2 \gg 6$, we can ignore 6 in comparison with x^2+y^2. Therefore, $y' \approx -\frac{x}{y}$ (refer to Exercise 13) and the curves look like circles.

3.2 Symmetry of Slope Fields

1. The slope field is symmetric across the x- and y-axes, and about origin. For example, to show symmetry across the y-axis, we replace x with $-x$ in $y^3\frac{dy}{dx} = x^3$, and find $-y^3\frac{dy}{dx} = -x^3$, which is equivalent to $y^3\frac{dy}{dx} = x^3$.

3. (a) Replacing y with $-y$ gives $d(-y)/dx = (-y)/(2x)$ or $dy/dx = y/(2x)$ so we have symmetry across the x-axis. Replacing x with $-x$ gives $dy/d(-x) = y/(-2x)$ or $dy/dx = y/(2x)$ so we have symmetry across the y-axis. Replacing x with $-x$ and y with $-y$ gives $d(-y)/d(-x) = (-y)/(-2x)$ or $dy/dx = y/(2x)$ so we have symmetry about the origin.

(b) If $y = \sqrt{x}$, $y' = 1/(2\sqrt{x})$, so $y' - y/(2x) = 1/(2\sqrt{x}) - \sqrt{x}/(2x) = 0$. This curve passes through the point $(1,1)$ because $1 = \sqrt{1}$.

(c) The function $y = \sqrt{x}$ is only defined for $x \geq 0$ and $y \geq 0$, so it is impossible for this function to possess symmetry across either axis or about the origin.

5. If $y' = g(x,y)$ is symmetric across the y-axis, and $y(0)$ is defined, then the solution curve through $y(0)$ is an even function.

7. (a) See Figure 3-4.

(c) See Figure 3-5.

(e) See Figure 3-6.

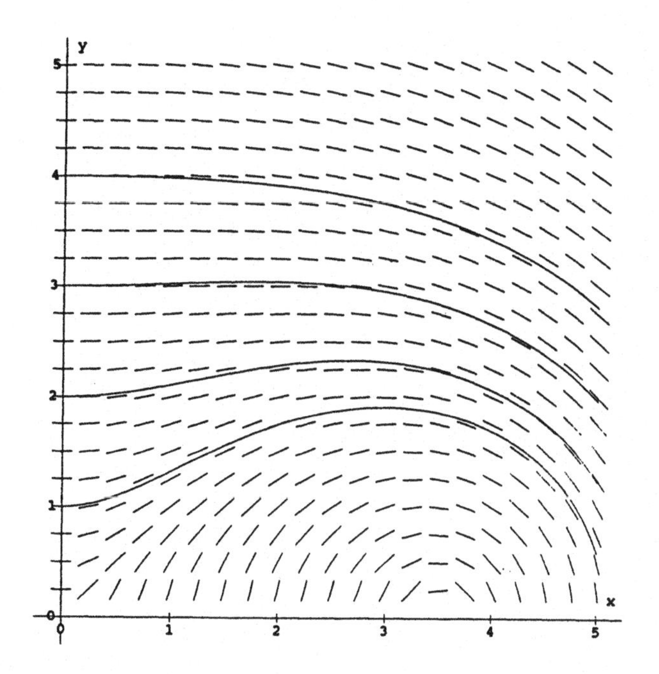

Figure 3-4 Exercise 7(a).

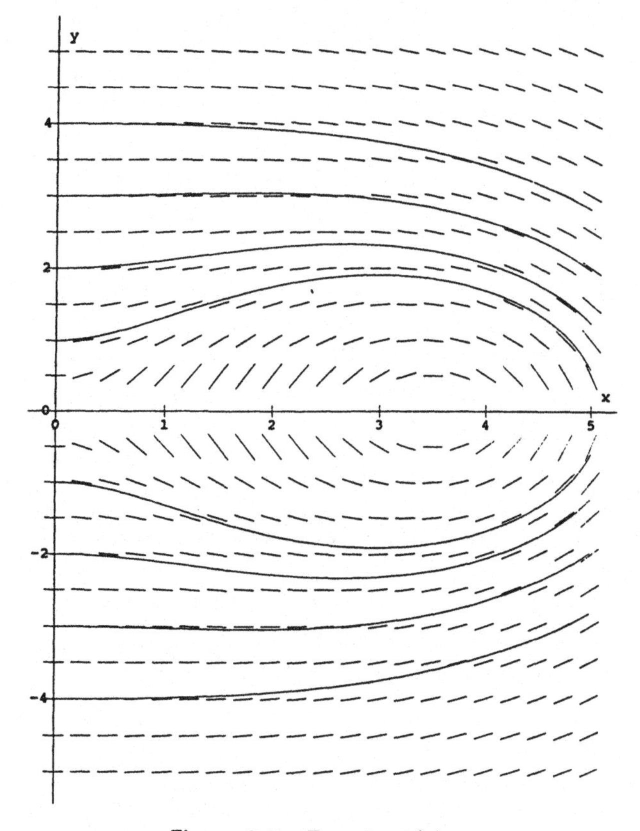

Figure 3-5 Exercise 7(c).

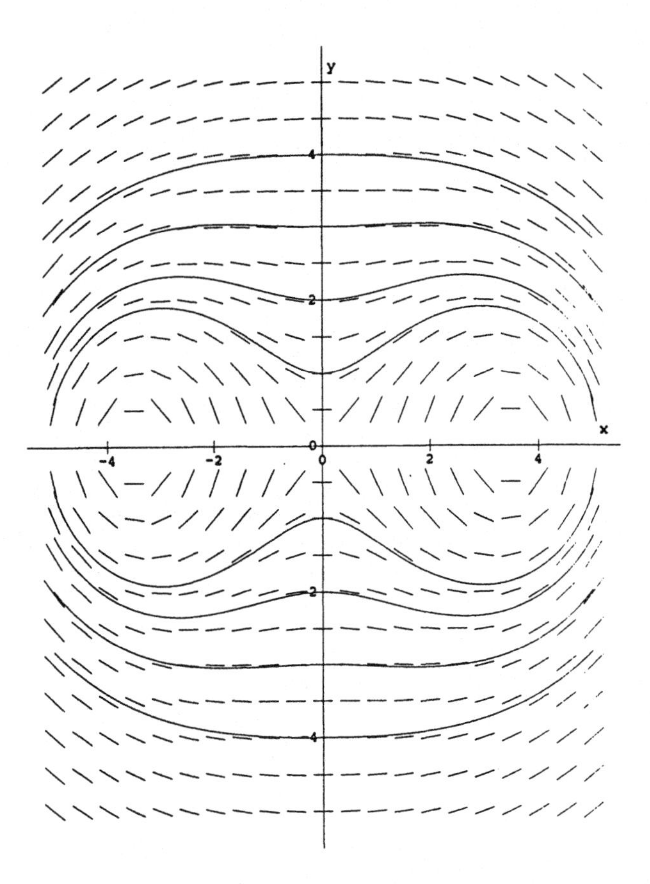

Figure 3-6 Exercise 7(e).

9. **(a)** If $y' = f(x)g(y)$ is symmetric across the x-axis, then replacing y with $-y$ leaves the differential equation unchanged. Thus $d(-y)/dx = f(x)g(-y)$ must be equivalent to $dy/dx = f(x)g(y)$. This requires that $g(-y) = -g(y)$, so $g(y)$ is an odd function.

(c) If $y' = f(x)g(y)$ is symmetric about the origin, then replacing x with $-x$ and y with $-y$ leaves the differential equation unchanged. Thus $d(-y)/d(-x) = f(-x)g(-y)$ must be equivalent to $dy/dx = f(x)g(y)$. This requires that $f(-x)g(-y) = f(x)g(y)$.

11. The differential equation of the type $y' = g(x)$ is invariant under the interchange of y with $y + C$ for every constant C because the derivative of C is zero. Therefore, whatever symmetries there are for $g(x)$ will hold when y is replaced by $y + C$. So, every member of the family of solutions differs by a vertical translation.

3.3 Numerical Solutions and Chaos

1. In Exercises (a) through (f), notice that Simpson's rule provides a much more accurate approximation than Euler's method.

(a) $y' = x^3$ subject to $y(1) = 1$.

i. Exact solution: $y(x) = \frac{1}{4}x^4 + \frac{3}{4}$.
$y(x_1) = 1.1160$, $y(x_2) = 1.2684$, $y(x_3) = 1.4640$, $y(x_4) = 1.7104$

ii. $y(x_1) = 1.1160$, $y(x_2) = 1.2684$, $y(x_3) = 1.4640$, $y(x_4) = 1.7104$

iii. $y(x_1) = 1.1000$, $y(x_2) = 1.2331$, $y(x_3) = 1.4059$, $y(x_4) = 1.6256$

iv. $y(x_1) = 1.0500$, $y(x_2) = 1.1079$, $y(x_3) = 1.1744$, $y(x_4) = 1.2505$, $y(x_5) = 1.3369$, $y(x_6) = 1.4345$, $y(x_7) = 1.5444$, $y(x_8) = 1.6674$

(c) $y' = e^{-x}$ subject to $y(0) = 1$.

i. Exact solution: $y(x) = -e^{-x} + 2$.
$y(x_1) = 1.0952$, $y(x_2) = 1.1813$, $y(x_3) = 1.2592$, $y(x_4) = 1.3297$

ii. $y(x_1) = 1.0952$, $y(x_2) = 1.1813$, $y(x_3) = 1.2592$, $y(x_4) = 1.3297$

iii. $y(x_1) = 1.1000$, $y(x_2) = 1.1905$, $y(x_3) = 1.2724$, $y(x_4) = 1.3464$

iv. $y(x_1) = 1.0500$, $y(x_2) = 1.0976$, $y(x_3) = 1.1428$, $y(x_4) = 1.1858$, $y(x_5) = 1.2268$, $y(x_6) = 1.2657$, $y(x_7) = 1.3028$, $y(x_8) = 1.3380$

(e) $y' = \frac{1}{1+x^2}$ subject to $y(1) = \frac{\pi}{4}$.

i. Exact solution: $y(x) = \arctan x$.
$y(x_1) = 0.8330$, $y(x_2) = 0.8761$, $y(x_3) = 0.9151$, $y(x_4) = 0.9505$

ii. $y(x_1) = 0.8330$, $y(x_2) = 0.8761$, $y(x_3) = 0.9151$, $y(x_4) = 0.9505$

iii. $y(x_1) = 0.8354$, $y(x_2) = 0.8806$, $y(x_3) = 0.9216$, $y(x_4) = 0.9588$

iv. $y(x_1) = 0.8104$, $y(x_2) = 0.8342$, $y(x_3) = 0.8568$, $y(x_4) = 0.8783$, $y(x_5) = 0.8988$, $y(x_6) = 0.9183$, $y(x_7) = 0.9369$, $y(x_8) = 0.9546$

3. **(i)** See Figure 3-7.

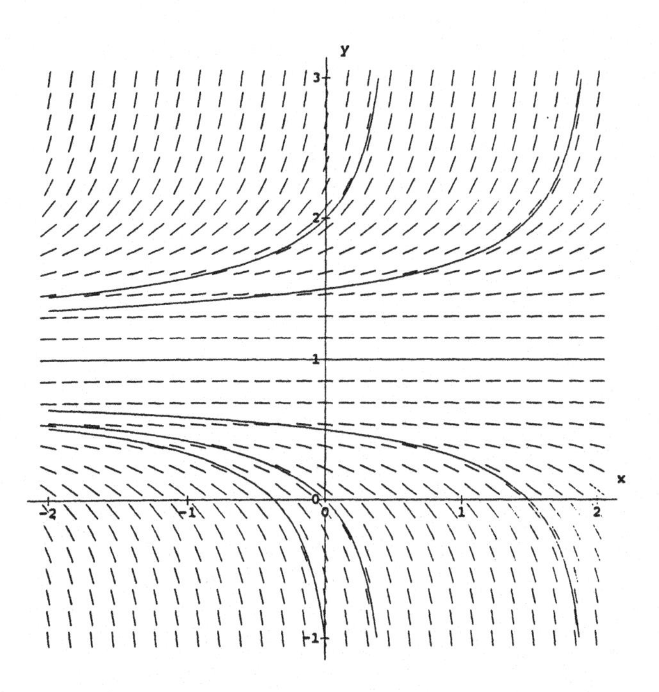

Figure 3-7 Exercise 3(i).

(iii) See Figure 3-8.

5. Because the numerical solution crosses the horizontal slope field at $y = 1$, the numerical solution and the slope field are not consistent. Because of the sharp changes in the numerical solution (which indicates a discontinuous derivative) the slope field is likely to be correct. It looks as though the step-size in the numerical method was set at about 0.3. Make the step-size smaller.

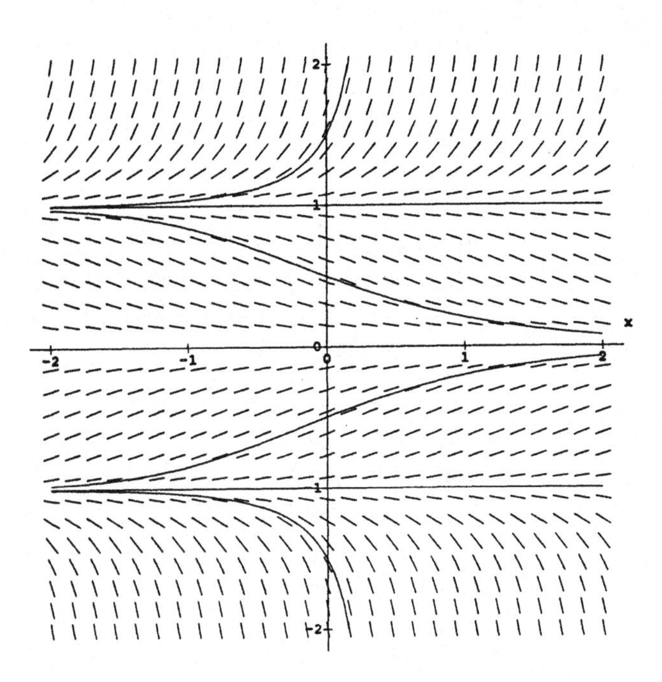

Figure 3-8 Exercise 3(iii).

7. **(a)** $y' = y - e$ is an autonomous equation with equilibrium solution $y(x) = e$, and nonequilibrium solutions

$$\int \frac{dy}{y-e} = \ln|y-e| = x + C,$$

or, $y - e = ce^x$, so $y(x) = e + ce^x$. The initial condition $y(0) = e - 1$ requires that $c = -1$, so $y(x) = e - e^x$, and $y(1) = 0$.

(b) With step-sizes of 0.1, 0.05, 0.25, and 0.125 we find $y(1) \approx 0.124539, 0.064984, 0.033218$, and 0.016797. Because the exact answer is 0, these values also represent the error made. A plot of the logarithm of the error as a function of the logarithm of h shows a straight line with slope 0.96, which is approximately 1. This is consistent with Euler's method being of first order.

(c) With step-sizes of 0.1, 0.05, 0.25, and 0.125 we find $y(1) \approx 0.004201, 0.001091, 0.000278$, and 0.000070. Because the exact answer is 0, these values also represent the error made. A plot of the logarithm of the error as a function of the logarithm of h shows a straight line with slope 1.97, which is approximately 2. This is consistent with Heun's method being of second order.

(d) With step-sizes of 0.1, 0.05, 0.25, and 0.125 we find $y(1) \approx 0.0000020843, 0.0000001358, 0.0000000087$, and 0.0000000005. Because the exact answer is 0, these values also represent the error made. A plot of the logarithm of the error as a function of the logarithm of h shows a straight line with slope 4.00. This is consistent with the Runge-Kutta 4 method being of fourth order.

9. Depending on the numerical method and the step-size, all values seem to get very large near $x = 1$, and never reach $x = 2$. The problem is due to $y(x)$ having a vertical asymptote somewhere near $x = 1$.

11. With the step-size h, for $h = 0.18$ to $h \approx 0.282$ we find period 1 solutions. For $h \approx 0.282$ to $h \approx 0.301$ we find period 2 solutions. For $h \approx 0.301$ to $h \approx 0.310$ we find period

4 solutions. For $h = 0.18$ to $h \approx 0.20$ all period 1 solutions have the same ultimate behavior — namely, 1. For $h \approx 0.20$ to $h \approx 0.282$, the ultimate behavior depends on h and decreases with h. For example, with $h = 0.28$ the ultimate behavior is about 0.5.

13. **(a)** Experimenting we find that if $y(0) < 0.8862$ all solutions tend to $-\infty$, whereas if $y(0) > 0.8863$ all solutions tend to ∞. Somewhere between 0.8862 and 0.8863 there is an initial condition whose solution approaches 0.

(b) Solutions will be increasing if $-1 + 2xy > 0$, that is, for $y > 1/(2x)$. The isocline of slope -1 is the x-axis. Thus, if a solution reaches the x-axis it decreases. The isocline of slope 0 is the hyperbola $1/(2x)$, and if a solution crosses this curve it does so horizontally and then increases.

(c) A small change in the initial condition — from $y(0) = 0.8862$ to $y(0) = 0.8863$ — results in a major change in the long-term behavior.

(d) $y(x) = \left[y_0 - \frac{\sqrt{\pi}}{2} erf(x)\right] e^{x^2}$, has the property that $y(0) = y_0$, so the initial condition is satisfied. Also,

$[erf(x)]' = \frac{2}{\sqrt{\pi}} e^{-x^2}$, so

$$y' = \left[y_0 - \frac{\sqrt{\pi}}{2} erf(x)\right]' e^{x^2} + \left[y_0 - \frac{\sqrt{\pi}}{2} erf(x)\right] \left(e^{x^2}\right)'$$
$$= \left(0 - \frac{\sqrt{\pi}}{2} \frac{2}{\sqrt{\pi}} e^{-x^2}\right) e^{x^2} + \left[y_0 - \frac{\sqrt{\pi}}{2} erf(x)\right] 2xe^{x^2}$$
$$= -1 + 2x \left[y_0 - \frac{\sqrt{\pi}}{2} erf(x)\right] e^{x^2}$$
$$= -1 + 2xy.$$

(e) As $x \to \infty$ we know $erf(x) \to 1$, so as $x \to \infty$ we have $y_0 - \frac{\sqrt{\pi}}{2} erf(x) \to y_0 - \frac{\sqrt{\pi}}{2}$.

If $y_0 - \frac{\sqrt{\pi}}{2} > 0$ then $y(x) = \left[y_0 - \frac{\sqrt{\pi}}{2} erf(x)\right] e^{x^2} \to \infty$, as $x \to \infty$.

If $y_0 - \frac{\sqrt{\pi}}{2} < 0$ then $y(x) = \left[y_0 - \frac{\sqrt{\pi}}{2} erf(x)\right] e^{x^2} \to -\infty$, as $x \to \infty$.

If $y_0 - \frac{\sqrt{\pi}}{2} = 0$ then $y(x) = \left[y_0 - \frac{\sqrt{\pi}}{2} erf(x)\right] e^{x^2} \to 0$, as $x \to \infty$.

Because $\frac{\sqrt{\pi}}{2} \approx 0.886226925$, this is consistent with what we found in part (a). The initial condition $y(0) = \frac{\sqrt{\pi}}{2}$ is the one which separates the behavior of the solutions.

3.4 Comparing Solutions of Differential Equations

1. If we mimic the procedure in Example 3.6 of the text, we have that

$y^2 < x^n + y^2 < 1 + y^2$ for $0 < x < 1$ and $n > 1$.

Now the initial value problem $y' = y^2$, $y(0) = 1$, has the explicit solution $y(x) = 1/(1 - x)$, while the initial value problem $y' = 1 + y^2$, $y(0) = 1$, has the explicit solution $y(x) = \tan(x + \pi/4)$. Because these two explicit solutions have asymptotes at $x = 1$ and $\pi/4$ respectively, by the Comparison Theorem, the solution of the initial value problem $y' = x^n + y^2$, $y(0) = 1$, is bounded between the solutions having these asymptotes, so it must have a vertical asymptote somewhere in this interval.

3. If $1 < y < B$, then $1 < \sqrt{y} < \sqrt{B}$, $1 < y + B$, and $1 < \sqrt{y} + \sqrt{B}$. If we multiply the last inequality by the positive number $\sqrt{B} - \sqrt{y}$ we have $\sqrt{B} - \sqrt{y} < B - y$. Similarly, multiplying $1 < y + B$ by the positive quantity $B - y$, gives $B - y < B^2 - y^2$. By the Comparison Theorem, this means that the solution of $n = 1$ is bounded below by the solution for $n = 1/2$ and above by that for $n = 2$.

5. (a) We complete the square by writing
$x^2 + y^2 - xy - 1 = x^2 - xy + (1/4)y^2 + (3/4)y^2 - 1$
$= [x - (1/2)y]^2 + (3/4)y^2 - 1$. This shows that
$(3/4)y^2 - 1 < x^2 + y^2 - xy - 1$.

(b) Comparing $y' = (3/4)y^2 - 1$ with Exercise 4, part (c), we have $a = \sqrt{3}/2$, $b = 1$, and $\alpha = \sqrt{3}$. Using the result from part (c) gives a vertical asymptote at $x = \ln 5/\sqrt{3}$. Because, $(3/4)y^2 - 1 < x^2 + y^2 - xy - 1$, from the Comparison Theorem we have that solutions of $y' = x^2 + y^2 - xy - 1$ will always be larger than those of $y' = (3/4)y^2 - 1$. Thus solutions of $y' = x^2 + y^2 - xy - 1$ will approach a vertical asymptote located to the left of $x = \ln 5/\sqrt{3}$.

3.5 Finding Power Series Solutions

1. (a) $y(x) = c_0 + c_1x + c_2x^2 + c_3x^3 + c_4x^4 + \cdots$.
$y'(x) = c_1 + 2c_2x + 3c_3x^2 + 4c_4x^3 + \cdots$.
$c_1 + 2c_2x + 3c_3x^2 + 4c_4x^3 + \cdots = -c_0 - c_1x - c_2x^2 - c_3x^3 + \cdots$.
$c_1 = -c_0$, $2c_2 = -c_1$, $3c_3 = -c_2$, $4c_4 = -c_3$, $\cdots$.
$c_1 = -c_0$,
$c_2 = -c_1/2 = c_0/2!$,
$c_3 = -c_2/3 = -c_0/3!$,
$c_4 = -c_3/4 = c_0/4!$, $\cdots$.
$y(x) = c_0 + c_1x + c_2x^2 + c_3x^3 + \cdots$
$= c_0 - c_0x + \frac{1}{2!}c_0x^2 - \frac{1}{3!}c_0x^3 + \frac{1}{4!}c_0x^4 + \cdots$
$= c_0\left(1 - x + \frac{1}{2!}x^2 - \frac{1}{3!}x^3 + \frac{1}{4!}x^4 + \cdots\right) = c_0e^{-x}$.

(c) $y(x) = c_0 + c_1x + c_2x^2 + c_3x^3 + c_4x^4 + \cdots$.
$y'(x) = c_1 + 2c_2x + 3c_3x^2 + 4c_4x^3 + \cdots$.
$c_1 + 2c_2x + 3c_3x^2 + 4c_4x^3 + \cdots + c_1x + 2c_2x^2 + 3c_3x^3 + 4c_4x^4 + \cdots$
$= 2c_0 + 2c_1x + 2c_2x^2 + 2c_3x^3 + \cdots$.
$c_1 = 2c_0$, $2c_2 = c_1$, $3c_3 = 0$, $4c_4 = -c_3$, $\cdots$.
$c_1 = 2c_0$,
$c_2 = c_1/2 = c_0$,
$c_3 = 0$,
$c_4 = -c_3/4 = 0$, $\cdots$.
$y(x) = c_0 + c_1x + c_2x^2 + c_3x^3 + \cdots$
$= c_0 + 2c_0x + c_0x^2 + 0x^3 + 0x^4 + \cdots$
$= c_0\left(1 + 2x + x^2\right) = c_0(1 + x)^2$.

3. $y(x) = c_0 + c_1x + c_2x^2 + c_3x^3 + c_4x^4 + c_5x^5 + c_6x^6 \cdots$.
$y'(x) = c_1 + 2c_2x + 3c_3x^2 + 4c_4x^3 + 5c_5x^4 + 6c_6x^5 + \cdots$.
$c_1 + 2c_2x + 3c_3x^2 + 4c_4x^3 + 5c_5x^4 + 6c_6x^5 + \cdots$
$= -2c_0x - 2c_1x^2 - 2c_2x^3 - 2c_3x^4 - 2c_4x^5 - 2c_5x^6 + \cdots$.
$c_1 = 0$, $2c_2 = -2c_0$, $3c_3 = -2c_1$, $4c_4 = -2c_2$, $5c_5 = -2c_3$, $6c_6 = -2c_4$, $\cdots$.
$c_1 = 0$,
$c_2 = -c_0$,

$$c_3 = -2c_1/3 = 0,$$
$$c_4 = -2c_2/4 = c_0/2!,$$
$$c_5 = -2c_3/5 = 0,$$
$$c_6 = -2c_4/6 = -c_0/3!, \cdots.$$
$$y(x) = c_0 + c_1 x + c_2 x^2 + c_3 x^3 + c_4 x^4 + c_5 x^5 + c_6 x^6 + \cdots$$
$$= c_0 - c_0 x^2 + \tfrac{1}{2!} c_0 x^4 - \tfrac{1}{3!} c_0 x^6 + \cdots$$
$$= c_0 \left(1 - x^2 + \tfrac{1}{2!} x^4 - \tfrac{1}{3!} x^6 + \cdots\right) = c_0 e^{-x^2}.$$

4. MODELS AND APPLICATIONS LEADING TO NEW TECHNIQUES

4.1 Solving Separable Differential Equations

1. **(a)** Separable

   (c) Neither

   (e) Neither

   (g) Autonomous and Separable

3. Integrate $yy' = x$ with respect to x to obtain $\frac{y^2}{2} = \frac{x^2}{2} + C$. $\sqrt{y^2} = \sqrt{x^2 + 2C}$.

   $y(x) = \pm\sqrt{x^2 + 2C}$.

5. **(a)** $y(x) = 0$ is an equilibrium solution. $y(e) = 1$ does not lie on this solution.

   Nonequilibrium solutions $\frac{1}{y}y' = \frac{1}{x}$. $\ln|y| = \ln|x| + C$, so

   $y(x) = \pm e^C x$. $y(e) = 1 \Rightarrow e^C = e^{-1}$, so

   $y(x) = x/e$, $x > 0$.

   (c) $y(x) = \pm 2$ are equilibrium solutions. $y(8) = 1$ does not lie on this solution.

   Nonequilibrium solutions $(4 - y^2)^{-1/2}y' = x^{-1}$.

   $\arcsin(y/2) = \ln|x| + C$.

   $y(x) = 2\sin(\ln|x| + C)$, $-\frac{\pi}{2} < \ln|x| + C < \frac{\pi}{2}$.

   $y(8) = 1 \Rightarrow \ln 8 + C = \frac{\pi}{6}$, so

   $y(x) = 2\sin\left(\ln\frac{x}{8} + \frac{\pi}{6}\right)$, where $-\frac{\pi}{2} < \ln\frac{x}{8} + \frac{\pi}{6} < \frac{\pi}{2}$.

   (e) $y(x) = -2$ is an equilibrium solution.

   Nonequilibrium solutions $\frac{1}{y+2}y' = -\frac{8x+1}{4x^2+x-1}$.

   $\ln|y+2| = -\ln\left|4x^2 + x - 1\right| + C$.

   $y(x) = \frac{K}{4x^2+x-1} - 2$, $K \neq 0$.

   $y(1) = -2$ is on the equilibrium solution $y(x) = -2$.

   (g) $y(x) = 0$ is an equilibrium solution. $y(\pi/2) = 1$ does not lie on this solution.

   Nonequilibrium solutions $(y + 1/y)y' = -\sin x e^{-\cos x}$.

   $\frac{y^2}{2} + \ln|y| = C - e^{-\cos x}$.

   $y(\pi/2) = 1 \Rightarrow 1/2 = C - e^0$, so $C = 3/2$ and

   $\frac{y^2}{2} + \ln y = \frac{3}{2} - e^{-\cos x}$, $y > 0$.

   (i) No equilibrium solutions. Nonequilibrium solutions of

   $3y^2y' = \sin x^2$. $y^3 = \int_0^x \sin t^2\,dt + C$.

   $y(x) = \left(\int_0^x \sin t^2\,dt + C\right)^{1/3}$

   $y(0) = 3$ requires $3 = C^{1/3}$, so $C = 27$ and

   $y(x) = \left(\int_0^x \sin t^2\,dt + 27\right)^{1/3}$

(k) $y = 0$ is an equilibrium solution. $y(0) = 2$ does not lie on an equilibrium solution.
Nonequilibrium solutions of $y'/y = 2x/(x^2 - 1)$.
$\ln|y| = \ln|x^2 - 1| + C$.
$y(x) = K(x^2 - 1)$, $K \neq 0$.
$y(0) = 2$ requires $2 = -K$, so
$y(x) = 2(1 - x^2)$, where $-1 < x < 1$.

7. **(a)** Let $y(x)$ be the number of pounds of salt in the container at time x in minutes.
$y' = 3 \cdot 4 - 4y/200 = 12 - y/50$. $y(0) = 0$.
$y(x) = 600$ is an equilibrium solution. $y(0) = 0$ does not lie on this solution.
Nonequilibrium solutions
$y'/(600 - y) = 1/50$.
$-\ln|600 - y| = x/50 + C$.
On physical grounds $600 - y \geq 0$, so $y(x) = 600 - e^{-x/50}e^{-C}$.
$y(0) = 0$, so $e^{-C} = 600$ and
$y(x) = 600\left(1 - e^{-x/50}\right)$.

(b) A concentration of 2 pounds per gallon occurs when there are 400 pounds of salt in the 200 gallon container. This will happen when
$400 = 600\left(1 - e^{-x/50}\right)$, or
$400/600 = 1 - e^{-x/50}$, so
$x = 50\ln 3 \approx 54.9306$ minutes ≈ 54 minutes 58.8 seconds.

(c) The concentration in the container at time x is given by $y(x)/200 = 3\left(1 - e^{-x/50}\right)$ pounds per gallon. This approaches 3 pounds per gallon for large values of time, because $3\left(1 - e^{-x/50}\right) \to 3$ as $x \to \infty$.
Intuitively, if 3 pounds of salt per gallon is added for a large amount of time to a container with initially no salt, as is the case in this exercise, you would expect the concentration in the container to eventually reach the concentration of 3 pounds of salt per gallon. Thus, the analytical solution agrees with intuition.

9. $P = 0$ is an equilibrium solution.
Nonequilibrium solutions of $(1/P)dP/dT = a/T^2$.
$\ln|P| = -a/T + C$. On physical grounds $P > 0$, so
$\ln P = -a/T + C$.
$P(T) = \exp(-a/T + C)$.
Letting $y = \ln P$ and $x = 1/T$, gives $y = -ax + C$. If plotting the data as $\ln P$ versus $1/T$ gives a straight line, the data is consistent with solutions of the Clapeyron equation with $-a$ equal to the slope and C as the vertical intercept. Doing this shows a straight line with slope -4458.965 and vertical intercept 18.276. The data falls nicely on the solution
$P(T) = \exp(-4458.965/T + 18.276)$.

11. You should find a circle.

13. **(a)** $y = 0$ is an equilibrium solution.
Nonequilibrium solutions of $(1 - 1/y)y' = 1/x - 1$.
$y - \ln|y| = \ln|x| - x + C$.
$y(1) = 3.5$ implies $3.5 - \ln(3.5) = -1 + C$, $C = 4.5 - \ln(3.5)$, so
$y - \ln y = \ln x - x + 4.5 - \ln(3.5)$, or
$x - 1 - \ln x = \ln y - y + 3.5 - \ln(3.5)$ for $x > 0$, $y > 0$. This gives an implicit solution as
$e^{x-1}/x = ye^{-y+3.5}/3.5$.

(b) Completing Figure 4.14 of the text should result in an oval region.

4.2 Solving Differential Equations with Homogeneous Coefficients

1. Isoclines of slope m are given by $y/x + \sqrt{1+(y/x)^2} = m$, or

$m - y/x = \sqrt{1+(y/x)^2}$. Squaring both sides gives

$m^2 - 2my/x + (y/x)^2 = 1 + (y/x)^2$, or

$m^2 - 2my/x = 1$. This is the straight line

$y = \left[(m^2-1)/(2m)\right] x$.

3. No equilibrium solutions.

To find nonequilibrium solutions of $y' = y/x - \sqrt{1+(y/x)^2})$, put $z = y/x$, so

$y' = z + xz' = z - \sqrt{1+z^2}$, or

$z'/\sqrt{1+z^2} = -1/x$. Thus

$\ln\left(z+\sqrt{1+z^2}\right) = -\ln(-x) + C$.

$z + \sqrt{1+z^2} = -e^C/x$.

$xz + x\sqrt{1+z^2} = -e^C$. Because $z = y/x$ and $x = -\sqrt{x^2}$ (remember $x < 0$)

$y - \sqrt{x^2+y^2} = -e^C$.

$-y + \sqrt{y^2+x^2} = e^C = K > 0$.

$y^2 + x^2 = (K+y)^2 = K^2 + 2Ky + y^2$.

$y = (x^2 - K^2)/(2K) = x^2/(2K) - K/2$. Now let $1/K = c$.

5. **(a)** $(x^2+3y^2)y' + 2xy = 0$.

$$y' = -\frac{2xy}{x^2+3y^2} = -\frac{2\left(\frac{y}{x}\right)}{1+3\left(\frac{y}{x}\right)^2}.$$

A differential equation with homogeneous coefficients.
Put $y = xz$ to find

$$xz' + z = -\frac{2z}{1+3z^2}.$$

$$xz' = -\frac{2z}{1+3z^2} - z = -\frac{2z+z+3z^3}{1+3z^2} = -\frac{3z+3z^3}{1+3z^2}.$$

The equilibrium solution is $z(x) = 0$, that is, $y(x) = 0$.
To find the nonequilibrium solutions:

$$\frac{1+3z^2}{3z+3z^3} z' = -\frac{1}{x}.$$

$$\int \frac{1+3z^2}{3z+3z^3}\, dz = -\int \frac{1}{x}\, dx.$$

To evaluate the first integral, put $u = 3z + 3z^3$ so $du = 3\left(1+3z^2\right) dz$.

$$\int \frac{1}{3u}\, du = -\ln|x| + C.$$

$$\frac{1}{3}\ln|u| = -\ln|x| + C.$$

$\ln\left|ux^3\right| = 3C$.

$ux^3 = A$, where $A = \pm e^{3C} \neq 0$.
$\left(3z + 3z^3\right) x^3 = A$.
$3\left[\left(\frac{y}{x}\right) + \left(\frac{y}{x}\right)^3\right] x^3 = A$.
$x^2y + y^3 = \frac{1}{3}A$.

(c) $(x^2 - xy + y^2)y' + y^2 = 0$.

$$y' = -\frac{y^2}{x^2 - xy + y^2} = -\frac{(\frac{y}{x})^2}{1 - \frac{y}{x} + (\frac{y}{x})^2}.$$

A differential equation with homogeneous coefficients.
Put $y = xz$ to find

$$xz' + z = -\frac{z^2}{1 - z + z^2}.$$

$$xz' = -\frac{z^2}{1 - z + z^2} - z = -\frac{z^2 + z - z^2 + z^3}{1 - z + z^2} = -\frac{z + z^3}{1 - z + z^2}.$$

The equilibrium solution is $z(x) = 0$, that is, $y(x) = 0$.
To find the nonequilibrium solutions:

$$\frac{1 - z + z^2}{z + z^3} z' = -\frac{1}{x}.$$

$$\int \frac{1 - z + z^2}{z + z^3}\, dz = -\int \frac{1}{x}\, dx = -\ln|x| + C.$$

To evaluate the remaining integral use partial fractions.

$$\frac{1 - z + z^2}{z + z^3} = \frac{1 - z + z^2}{z\,(1 + z^2)} = \frac{A}{z} + \frac{Bz + C}{1 + z^2},$$

so $1 - z + z^2 = A\left(1 + z^2\right) + (Bz + C)\,z = A + Cz + (A + B)\,z^2$. This leads to $A = 1$, $B = 0$, and $C = -1$, so

$$\int \frac{1 - z + z^2}{z + z^3}\, dz = \int \left(\frac{1}{z} - \frac{1}{1 + z^2}\right) dz = \ln|z| - \arctan z.$$

$$\ln|z| - \arctan z = -\ln|x| + C.$$

$$\ln|y| - \arctan\left(\frac{y}{x}\right) = C.$$

(e) $(xy - x^2)y' - y^2 = 0$.

$$y' = \frac{y^2}{xy - x^2} = \frac{\left(\frac{y}{x}\right)^2}{\frac{y}{x} - 1}.$$

A differential equation with homogeneous coefficients.
Put $y = xz$ so that $xz' + z = z^2/(z - 1)$, or

$$xz' = \frac{z}{z - 1}.$$

The equilibrium solution is $z(x) = 0$, that is, $y(x) = 0$.
To find nonequilibrium solutions:

$$\int \frac{z - 1}{z}\, dz = \int \frac{1}{x}\, dx.$$

$$z - \ln|z| = \ln|x| + C.$$

$y/x - \ln|y/x| = \ln|x| + C$.
$y = x(\ln|y| + C)$.

(g) $x^2y' + y^2 - xy = 0$.

$$y' = \frac{xy - y^2}{x^2} = \frac{\frac{y}{x} - (\frac{y}{x})^2}{1}.$$

A differential equation with homogeneous coefficients.
Put $y = xz$, so that $xz' + z = z - z^2$, or

$$xz' = -z^2.$$

The equilibrium solution is $z(x) = 0$, that is, $y(x) = 0$.
To find the nonequilibrium solutions:

$$\int -\frac{1}{z^2}\,dz = \int \frac{1}{x}\,dx.$$

$$\frac{1}{z} = \ln|x| + C.$$

$$\frac{x}{y} = \ln|x| + C.$$

$$y = \frac{x}{\ln|x| + C}.$$

(i) $2xyy' + x^2 + y^2 = 0$.

$$y' = -\frac{x^2 + y^2}{2xy} = -\frac{x}{2y} - \frac{y}{2x}.$$

A differential equation with homogeneous coefficients.
Put $y = xz$, so that $xz' + z = -1/(2z) - z/2$, or

$$xz' = \frac{-1 - 3z^2}{2z}.$$

There are no equilibrium solutions.
To find nonequilibrium solutions:

$$\int \frac{2z}{1 + 3z^2}\,dz = -\int \frac{1}{x}\,dx.$$

$$\frac{1}{3}\ln(1 + 3z^2) = -\ln|x| + C,$$

or $x^3 + 3xy^2 = c$.

(k) $xy' - y\ln(y/x) - y = 0$.

$$y' = \left[\ln\left(\frac{y}{x}\right) + 1\right]\left(\frac{y}{x}\right).$$

A differential equation with homogeneous coefficients.
Put $y = xz$, so that $xz' + z = (\ln z + 1)z$, or

$$xz' = z\ln z.$$

There are no equilibrium solutions.
To find nonequilibrium solutions:

$$\int \frac{1}{z\ln z}\,dz = \int \frac{1}{x}\,dx.$$

$$\ln|\ln z| = \ln|x| + C.$$

$\ln z = cx$, where $c = \pm e^C$
$z = e^{cx}$, or
$y = xe^{cx}$.

7. (a) $F(x,y) = x^2 + 3xy - x^3(x+y)^{-1}$ is a homogeneous function of degree 2.

$F(tx,ty) = (tx)^2 + 3(tx)(ty) - (tx)^3(tx+ty)^{-1}$
$= t^2x^2 + 3t^2xy - t^3x^3t^{-1}(x+y)^{-1}$
$= t^2x^2 + 3t^2xy - t^2x^3(x+y)^{-1}$
$= t^2\left[x^2 + 3xy - x^3(x+y)^{-1}\right].$
$= t^2F(x,y).$

(c) $F(x,y) = x\sin(\frac{x}{y}) + \frac{x^2}{y}$ is a homogeneous function of degree 1.

$F(tx,ty) = tx\sin\left(\frac{tx}{ty}\right) + \frac{(tx)^2}{ty}$
$= tx\sin\left(\frac{x}{y}\right) + \frac{t^2x^2}{ty}$
$= t\left[x\sin\left(\frac{x}{y}\right) + \frac{x^2}{y}\right]$
$= tF(x,y).$

(e) $F(x,y) = \ln x - \ln y + e^{\frac{x}{y}}$ is a homogeneous function of degree 0.

$F(tx,ty) = \ln(tx) - \ln(ty) + e^{\frac{tx}{ty}}$
$= \ln t + \ln x - \ln t - \ln y + e^{\frac{x}{y}}$
$= \ln x - \ln y + e^{\frac{x}{y}}$
$= F(x,y).$

9. (a) $y' = -(x+y-2)/(x-y+4)$.

$a_1b_2 - a_2b_1 = -2$, so let
$x = u + \alpha$, $y = v + \beta$.
$x + y - 2 = u + v + \alpha + \beta - 2$
$x - y + 4 = u - v + \alpha - \beta + 4.$
$\alpha + \beta - 2 = 0$, $\alpha - \beta + 4 = 0$ requires
$\alpha = -1$ $\beta = 3$, so $x = u - 1$, $y = v + 3$ and the differential equation becomes
$dv/du = -(u+v)/(u-v) = -(1+v/u)/(1-v/u).$
Put $v = uz$, so that $dv/du = z + udz/du = -(1+z)/(1-z)$, or

$$u\frac{dz}{du} = -\frac{1+2z-z^2}{1-z}.$$

The equilibrium solutions are $z(x) = 1 \pm \sqrt{2}$, that is, $y(x) = x\left(1 \pm \sqrt{2}\right) + 4 \pm \sqrt{2}$.
To find nonequilibrium solutions:

$$\int \frac{1-z}{1+2z-z^2}\,dz = -\int \frac{1}{u}\,du.$$

$$\frac{1}{2}\ln\left|1+2z-z^2\right| = -\ln|u| + C,$$

or $u^2 + 2u^2z - u^2z^2 = c$, where $c = \pm e^{2C}$.
$u^2 + 2uv - v^2 = c$, or
$(x+1)^2 + 2(x+1)(y-3) - (y-3)^2 = c.$

(c) $y' = (x - 2y + 1)/(2x - 4y + 3)$.

$a_1b_2 - a_2b_1 = 0$, so let
$z = x - 2y.$
$y' = (-z' + 1)/2 = (z+1)/(2z+3).$

$$\frac{dz}{dx} = \frac{1}{2z+3}.$$

No equilibrium solutions.

To find nonequilibrium solutions:

$$\int (2z+3)\,dz = \int 1\,dx.$$

$$z^2 + 3z = x + C,$$

or $(x-2y)^2 + 3(x-2y) = x + C$.

$(x-2y)^2 + 2x - 6y = C$.

4.3 Models — Deriving Differential Equations From Data

1. **(a)** Using central differences, the best fit of dT/dt to $\alpha t + \beta$ gives $\alpha \approx 0.0875$ and $\beta \approx -2.465$. α is positive.

Integration gives $T(t) = \alpha t^2/2 + \beta t + c$. Because $T(0) = 82.3$, $c = 82.3$.

As $t \to \infty$, $T \to \infty$, because α is positive. This is not a reasonable model for the cooling of coffee.

(b) Using central differences, the best fit of dT/dt to $aT + b$ gives $a \approx -0.0434$, $b \approx 1.066$.

Integration gives $T(t) = \left(Ce^{at} - b\right)/a$, where $C = b + 82.3a = -2.506$. a is negative.

As $t \to \infty$, $T \to -\frac{b}{a}$, which is a positive number (the ambient temperature). Yes, this is a reasonable model for the cooling of coffee.

(c) $T(t) = \left(Ce^{at} - b\right)/a = \left[C\sum_{k=0}^{k=\infty}(at)^k/k! - b\right]/a$

$= (C-b)/a + (C/a)(at) + (C/a)(at)^2/2! + (C/a)(at)^3/3! + \cdots$

$= (C-b)/a + Ct + Cat^2/2 + (C/a)(at)^3/3! + \cdots$.

This agrees with the expression in part (a) if $(C-b)/a = c = 82.3$, $C = \beta$, $Ca = \alpha$. The first of these three equations is exact, with the other numbers being close to being equal ($-2.506 \approx -2.465$, $0.1088 \approx 0.0875$).

3. **(a)** The phase line for the first time period will show a stable equilibrium at 70. The phase line for the second time period will have a stable equilibrium at 400. The solution curve will start at 180, be concave down and decreasing until it reaches 120, then be increasing and concave up until it reaches 180 again.

(b) Let $T(t)$ be the temperature of the coffee at time t in minutes, measured from noon.

$T(t) = T_a + Ce^{kt}$, where k is negative. For the first thirty minutes we have $T_a = 70$ and $T(0) = 180$. For $t > 30$, we have $T_a = 400$ and $T(30) = 120$. Thus,

$T(t) = 70 + 110e^{kt}$ for $0 < t < 30$, and

$T(t) = 400 - 280e^{k(t-30)}$ for $30 < t$.

To determine the value of k, we use the fact that $T(30) = 120$ in our solution $T(t) = 70 + 110e^{kt}$. This gives $120 = 70 + 110e^{30k}$, $e^{30k} = 50/110$, and $k = (1/30)\ln(5/11) \approx -0.0263$. Thus, we have

$T(t) = 70 + 110e^{-0.0263t}$ for $0 < t < 30$. For $t > 30$, we have $T(t) = 400 - 280e^{-0.0263(t-30)}$.

(c) $T(t) = 180$ when $180 = 400 - 280e^{-0.0263(t-30)}$, giving $e^{-0.0263(t-30)} = 220/280$. From this we have $t = 30 + \ln(22/28)/(-0.0263) \approx 39.18$ minutes = 39 minutes 10 seconds after noon.

5. If, in $1/y(x) = [1/(bC)]\, e^{-abx} + 1/b$, we define $Y(x) = 1/y(x)$, $A = 1/(bc)$, $B = -ab$, and $C = 1/b$, we have $Y(x) = Ae^{Bx} + C$ where $B < 0$, which is the equation used in Exercise 4.

7. **(a)** $dL/dt = a + kL$.

The equilibrium solution is $L(t) = -a/k$. If $-a/k = L_e$, then

$dL/dt = a - aL/L_e = a(1 - L/L_e)$.

This is a reasonable model in that the growth rate is large for a small fish and decreases as the fish becomes longer.

L_e is the maximum length of the fish while a must be a positive number.

(b) This equation is similar to that for Newton's Law of Heating.

(c) Nonequilibrium solutions are obtained from

$$\int \frac{1}{1 - L/L_e}\, dL = \int a\, dt.$$

$-L_e \ln|1 - L/L_e| = at + C$.

$L(t) = L_e\left(1 - ce^{-at/L_e}\right)$, where $c = \pm e^C$.

There are three parameters in this solution, c, a, and L_e.

After some trial and error, the estimate of $L_e = 212$ makes the graph of $\ln(212 - L)$ versus t appear linear, with slope -0.054 and intercept 5.246. Thus,

$L(t) = 212 - \exp(5.246 - 0.054t)$, which agrees remarkably well with the data.

9. **(a)** DuLong-Petit Law, $T' = k(T - T_a)^n$ has solutions which are increasing if $T < T_a$ and decreasing if $T > T_a$, so $T(t) = T_a$ is a stable equilibrium.

Newton-Stefan Law, $T' = (T - T_a)\,[k + h\,(T - T_a)]$ has two equilibrium solutions, $T(t) = T_a$, and $T(t) = T_a - k/h < T_a$, because both k and h are negative. Using the Derivative Test for Stable or Unstable Equilibrium Solutions we find that $T(t) = T_a$ is stable, while $T(t) = T_a - k/h$ is unstable.

These models are similar to that of Newton's Law of Cooling because the equilibrium solution, $T(t) = T_a$ is stable.

(b) $T' = k(T - T_a)^n$.

Nonequilibrium solutions are obtained from

$$\int \frac{1}{(T - T_a)^n}\, dT = \int k\, dt.$$

$$\frac{(T - T_a)^{-n+1}}{-n+1} = kt + C,$$

or

$$T(t) = T_a + [(-n+1)(kt + C)]^{1/(1-n)}.$$

$T' = k(T - T_a) + h(T - T_a)^2$.

Nonequilibrium solutions are obtained from

$$\int \frac{1}{k(T - T_a) + h(T - T_a)^2}\, dT = \int 1\, dt.$$

$\ln|T - T_a| - \ln|k + h(T - T_a)| = kt + C$.

$(T - T_a)/[k + h(T - Ta)] = ce^{kt}$, where $c = \pm e^C$.

$T(t) = T_a + kc/\left(e^{-kt} - ch\right)$.

11. From $T' = k(T - T_a)$,

$T(t) = 72 + (98.6 - 72)e^{kt}$.

The temperature of the body must be greater than 72 degrees, so after 24 hours, $72.72 < 72 + 26.6e^{24k}$.

Thus $k < \ln(0.72/26.6)/24 \approx 0.1504$.

13. $T(t) = a + bt$,

$[T(t+h) - T(t)]/h = [a + b(t+h) - a - bt]/h = b$, giving

$dT/dt = [T(t+h) - T(t)]/h$.

Thus, using the right-hand quotient to calculate the derivative of a linear function gives an exact answer, while for a nonlinear function it will not be exact.

15. $T(t) = ce^{at}$,

$[\ln T(t+h) - \ln T(t)]/h = \left[\ln ce^{a(t+h)} - \ln ce^{at}\right]/h = \ln e^{ah}/h = a$, giving

$1/T(t)dT/dt = [\ln T(t+h) - \ln T(t)]/h$.

Thus, using the right-hand quotient or the central difference quotient to calculate the derivative of an exponential function will not give an exact answer.

17. $T(t) = c + ae^{bt}$,

$T'(t) = abe^{bt}$,

$\frac{\Delta_c T}{\Delta t} = [T(t+h) - T(t-h)]/(2h) =$

$\left[c + ae^{b(t+h)} - (c + ae^{b(t-h)})\right]/(2h) = ae^{at}\left(e^{bh} - e^{-bh}\right)/(2h)$. Thus,

$\frac{1}{T'(t)}\frac{\Delta_c T}{\Delta t} = \left(e^{bh} - e^{-bh}\right)/(2bh)$.

With $bT_H = -\ln 2$ and $2h = \lambda T_H$, we have $bh = -(\lambda/2)\ln 2$, so

$\frac{1}{T'(t)}\frac{\Delta_c T}{\Delta t} = \left(e^{-(\lambda/2)\ln 2} - e^{(\lambda/2)\ln 2}\right)/(-\lambda \ln 2)$.

If $\lambda = 1$ then

$\frac{1}{T'(t)}\frac{\Delta_c T}{\Delta t} = \left(e^{-(1/2)\ln 2} - e^{(1/2)\ln 2}\right)/(-\ln 2) \approx 1.02$, that is

$\frac{\Delta_c T}{\Delta t} \approx 1.02T'(t)$, so

$\frac{\Delta_c T}{\Delta t} - T'(t) \approx 0.02T'(t)$.

4.4 Models — Objects in Motion

1. From (4.50), $v(t) = V(e^{\alpha t} - 1)/(e^{\alpha t} + 1)$, where $V = 180$ and $\alpha = 2(32.2)/180 \approx 0.358$.

If T is the time when she reaches 90% of her terminal velocity V, then $v(T) = 0.9V$, so that $(e^{\alpha T} - 1)/(e^{\alpha T} + 1) = 0.9$. Solving for $e^{\alpha T}$ gives $0.1e^{\alpha T} = 1.9$, so $e^{\alpha T} = 19$, and $T = (1/\alpha)\ln(19) \approx 8.23$ sec.

From (4.52) we have $x(t) = (2V/\alpha)\ln\left[(e^{\alpha t} + 1)/2\right] - tV$.

At $t = T \approx 8.23$, she has fallen $x(T) = (2V/\alpha)\ln\left[(e^{\alpha T} + 1)/2\right] - TV = (2V/\alpha)\ln 10 - TV \approx 835$ feet.

3. **(a)** Note the phase line for the first time period has a stable equilibrium at $v = 180$, while for the second time period this stable equilibrium is at $v = 22$.

(b) From (4.50) and (4.52), until the parachute opens $v(t) = V(e^{\alpha t} - 1)/(e^{\alpha t} + 1)$ and, using $x(0) = 0$, $x(t) = (2V/\alpha)\ln[(e^{\alpha t} + 1)/2] - tV$, where $V = 180$, and $\alpha = 2(32.2)/180 \approx 0.358$.

The parachute opens when $x(t) = 9000$, that is, when

$9000 = 1006.21 \ln\left[(e^{0.358t} + 1)/2\right] - 180t$.

A graphical or numerical solution gives $t \approx 48.22$ seconds.

$v(48.22) = 180(e^{0.358\times 48.22} - 1)/(e^{0.358\times 48.22} + 1) \approx 180$ ft/sec.

(c) From part (b), $t_0 = 48.22$ and $v_0 = 180$, while from Exercise 2, $x_0 = 9000$, $\alpha = 2g/V \approx 2.927$, $V = 22$, $\beta = (22 - 180)/(22 + 180) \approx -0.7822$, so

$v(t) = 22\left[e^{2.927(t-48.22)} + 0.7822\right] / \left[0.7822e^{2.927(t-48.22)} - 0.7822\right]$

$x(t) = 9000 + 15.03\ln\left[\left(e^{2.927(t-48.22)} - 0.7822\right)/0.2178\right] - 22(t - 48.22)$.

Using a graphical or numerical solution gives $x(t) = 10000$ when $t \approx 92.64$ sec, so $v(92.64) \approx 22$ ft/sec.

5. $mv' = -mg$,

$v(t) = -gt + C$,

$y(t) = -gt^2/2 + Ct$, because $y(0) = 0$.

$v(t) = 0$ when $t = C/g$.

$H = y(C/g) = C^2/(2g)$. The time in the air,

$T = 2C/g$, so

$H = (g/2)(C/g)^2 = gT^2/8$.

If $T = 1/3$, $H = 32.2/72$ ft ≈ 5.4 inches.

It will take ballet dancers with long feet longer than 1/3 second to jump high enough to point their feet, so will not be able to jump in time to such music.

7. Terminal velocity will occur at the equilibrium solution mg/k. Denote this terminal velocity by $V = mg/k$ so the differential equation is $v' = g(V - v)/V$.

Integration gives $\ln(V - v) = -gt/V + C$, (remember $v < V$).

For an object falling from rest we have $v(0) = 0$ so $C = \ln V$, and so

$\ln[(V - v)/V] = -gt/V$.

Solving for v gives $v(t) = V\left(1 - e^{-gt/V}\right)$. (Notice that as $t \to \infty$, $v \to V$, as expected.)

Because $x' = v$ integration yields $x(t) = Vt + V^2e^{-gt/V}/g + C$.

The initial condition $x(0) = 0$ gives $C = -V^2/g$ and so

$x(t) = Vt - V^2\left(1 - e^{-gt/V}\right)/g$.

This model does not fit the data set well for any choice of V and so it is not a good model.

9. $mv' = -mg$,

$v(t) = -gt + v_0$,

$x(t) = -gt^2/2 + v_0t$, because $x(0) = 0$.

$v(t) = 0$ when $t = v_0/g$.

Total time in the air is $2v_0/g$.

Maximum height $= x(v_0/g) = v_0^2/(2g)$.

$x(t) > 0.75v_0^2/(2g)$ for $-gt^2/2 + v_0t > 0.75v_0^2/(2g)$. Factoring gives

$(g/2)(t-(3/2)v_0/g)(t-(1/2)v_0/g) < 0$, so the mass is in the top 25% of the trajectory for $(1/2)v_0/g < t < (3/2)v_0/g$. This time interval of v_0/g is 50% of the total time in the air.

4.5 Application — Orthogonal Trajectories

1. **(a)** Differentiating $y = -2x + \lambda$ gives the differential equation for this family of curves $y' = -2$.

The differential equation for the orthogonal trajectories is $y' = 1/2$ with solution $y = x/2 + C$.

(c) Differentiating $y^2 = \lambda x$ gives $2yy' = \lambda$ where $\lambda = y^2/x$. Thus, the differential equation for this family of curves is $y' = y/(2x)$.

The differential equation for the orthogonal trajectories is $y' = -2x/y$ with solution $y^2 = -2x^2 + C$.

(e) Differentiating $x^2 - \lambda^2 y^2 = 16$ gives $2x - 2\lambda^2 yy' = 0$ where $\lambda^2 = \left(x^2 - 16\right)/y^2$. Thus, the differential equation for this family of curves is $y' = xy/\left(x^2 - 16\right)$.

The differential equation for the orthogonal trajectories is $y' = \left(16 - x^2\right)/(xy)$ with solution $y^2 = -x^2 + 32\ln|x| + C$.

(g) Differentiating $x^2 + ay^2 = \lambda^2$ gives $2x + 2ayy' = 0$. Thus, the differential equation for this family is $y' = -x/(ay)$.

The differential equation for the orthogonal trajectories is $y' = ay/x$, with solution $\ln|y| = a\ln|x| + c$, or $y(x) = x^a C$.

(i) Differentiating $e^{-x}y = \lambda$ gives $-e^{-x}y + e^{-x}y' = 0$. Thus, the differential equation for this family is $y' = y$.

The differential equation for the orthogonal trajectories is $y' = -1/y$, with solution $y^2/2 = -x + C$, or $y(x) = \pm\sqrt{2C - 2x}$.

(k) Differentiating $e^{-ax}y = \lambda$ gives $-ae^{-x}y + e^{-x}y' = 0$. Thus the differential equation for this family is $y' = ay$.

The differential equation for the orthogonal trajectories is $y' = -1/(ay)$, with solution $y^2/2 = -x/a + C$, or $y(x) = \pm\sqrt{2C - 2x/a}$.

(m) We use both sides of the equation as the exponent of e giving $e^y = x^3 + \lambda$, so solving for λ gives $\lambda = e^y - x^3$. Differentiating gives $0 = e^y y' - 3x^2$, so $y' = 3x^2 e^{-y}$ is the differential equation for this family of curves.

The differential equation for the orthogonal trajectories is $y' = -e^y/(3x^2)$ with solution $y = -\ln\left[C - 1/(3x)\right]$.

(o) Differentiating $e^x \cos y = \lambda$ gives $e^x \cos y - e^x \sin y y' = 0$, so the differential equation for this family of curves is $y' = \cos y/\sin y$.

The differential equation for the orthogonal trajectories is $y' = -\sin y/\cos y$ with solution $\ln|\sin y| = -x + C$.

3. **(a)** All the straight lines shown either have slope 1 or slope -1. They are either parallel or perpendicular.

(b) The curves are called self-orthogonal because the family of curves contains members that cross at right angles.

(c) Differentiating gives $2yy' = 2(x + \lambda)$, from which we obtain $yy' = \pm|y|$.

This means that the equation for the orthogonal trajectories is $y' = \pm 1$.

Solving gives $y = \pm(x + C)$ or $y^2 = (x + C)^2$, the same form as the original family of curves.

5. Differentiating $xy = b$ gives $xy' + y = 0$, so the differential equation for this family of curves is $y' = -y/x$.

The differential equation for the orthogonal trajectories is $y' = x/y$ with solution $y^2/2 - x^2/2 = C$, which is also a family of hyperbolas.

7. $\tan(\psi_1 - y) = \sin(\psi_1 - y) / \cos(\psi_1 - y) = (\sin\psi_1 \cos y - \sin y \cos\psi_1) / (\cos\psi_1 \cos y + \sin\psi_1 \sin y)$, and $\sin(\pm\pi/2) = \pm 1$ and $\cos(\pm\pi/2) = 0$, gives $\tan(\psi_1 \pm \pi/2) = -1/\tan\psi_1$.

   (a) If $r = f(\theta) = \lambda\cos\theta$, $r' = -\lambda\sin\theta$.
   The differential equation for the orthogonal trajectories is $r'/r = \cos\theta/\sin\theta$.

   (b) Integration gives $\ln|r| = \ln|\sin\theta| + C$, or
   $r(\theta) = c\sin\theta$, where $c = \pm e^C$.

9. $r = f(\theta) = \lambda(1 - \sin\theta)$ so $f' = -\lambda\cos\theta$.
   The differential equation for the orthogonal trajectories is
   $r'/r = (1 - \sin\theta)/\cos\theta = \cos\theta/(1 + \sin\theta)$. Integration gives
   $\ln|r| = \ln|1 + \sin\theta| + C$, or
   $r(\theta) = c(1 + \sin\theta)$, where $c = \pm e^C$.

11. $r = f(\theta) = \lambda\theta^{-1/2}$ so $f' = -\lambda\theta^{-3/2}/2$.
   The differential equation for the orthogonal trajectories is
   $r'/r = 2\theta$. Integration gives
   $\ln|r| = \theta^2 + C$, or
   $r(\theta) = ce^{\theta^2}$, where $c = \pm e^C$.

13. If we let $a = 1/(m^2 + \lambda)$ and $b = 1/(n^2 + \lambda)$ then the original family of curves is $ax^2 + by^2 = 1$. Differentiating this gives the differential equation for this family of curves $ax + byy' = 0$. We must eliminate λ from these two equations. From this last equation we have $a = -byy'/x$, which when substituted into $ax^2 + by^2 = 1$ gives $b = 1/[y(xy' - y)]$. Thus, $a = y/[x(xy' - y)]$. We thus have
   $m^2 + \lambda = -x(xy' - y)/y'$ and
   $n^2 + \lambda = y(xy' - y)$, which when subtracted gives
   $m^2 - n^2 = (xy' - y)(x/y' + y)$ so
   $(m^2 - n^2)y' = (xy' - y)(x + yy')$, which contains no λ.
   If in this equation we replace y' by $-1/y'$ we find the differential equation for the orthogonal trajectories, namely,
   $-(m^2 - n^2)/y' = (-x/y' - y)(x - y/y')$, or
   $(m^2 - n^2)y' = (x + yy')(xy' - y)$, which is the original equation. Thus, they are self-orthogonal.

4.6 Piecing Together Differential Equations

1. **(a)** Using (4.45) with $v(0) = 0$ $g = 32.2$, and $k/m = 0.000972$ along with its solution from (4.50), with $\alpha = 0.354$, $V = 180$ gives
   $v(t) = 180(e^{0.354t} - 1)/(e^{0.354t} + 1)$, $0 < t < 6$.
   This gives $v(6) = 143$.
   For $t > 6$, we solve $v' = 32.2 - 1.464v$ with $v(6) = 143$.
   This gives
   $1/(-1.464)\ln(32.2 - 1.464v) = t + c$, or

$v(t) = 22 - Ke^{-1.464(t-6)}$.

$v(6) = 22 - K = 143$, giving $K = -121$ and

$v(t) = 22 + 121e^{-1.464(t-6)}$, $t > 6$.

For $0 < t < 6$ the graph of $v(t)$ versus t is increasing and concave down. This is consistent with $v' = g - (k/m)v^2$ and $v'' = -2(k/m)vv'$ because $g - (k/m)v^2 > 0$ and $v > 0$.

For $t > 6$ the graph of $v(t)$ versus t is decreasing and concave up. This is consistent with $v' = g - (b/m)v$ and $v'' = -(b/m)v'$ because $g - (b/m)v < 0$.

(b) If the parachute had never opened, the terminal velocity would have been 180 ft/sec, whereas, in part (a), it is 22 ft/sec.

3. For every interval, the solution of $y' = -ay$ is $y(t) = Ke^{-at}$.

For $0 < t < T$, we have the initial condition $y(0) = c$, giving $K = c$ and

$y(t) = ce^{-at}$, $0 < t < T$. Also, $\lim_{t \to T} y(t) = ce^{-aT}$.

For $T < t < 2T$, we have the initial condition $y(T) = c + ce^{-aT}$ giving $K = c\left(1 + e^{aT}\right)$ and

$y(t) = c\left(1 + e^{aT}\right)e^{-at}$, $T < t < 2T$. Also, $\lim_{t \to 2T} y(t) = c\left(e^{-2aT} + e^{-aT}\right)$.

For $2T < t < 3T$, we have the initial condition $y(2T) = c + c\left(e^{-2aT} + e^{-aT}\right)$ giving $K = c\left(1 + e^{aT} + e^{2aT}\right)$ and

$y(t) = c\left(1 + e^{aT} + e^{2aT}\right)e^{-at}$, $2T < t < 3T$. Also, $\lim_{t \to 3T} y(t) = c\left(e^{-3aT} + e^{-2aT} + e^{-aT} + 1\right)]$

Continuing in this manner gives

$y(t) = c\left(\sum_{n=0}^{n=N} e^{naT}\right)e^{-at}$, $NT < t < (N+1)T$, which using the sum of $n+1$ terms in a geometric series becomes

$y(t) = c\left[1 - e^{(N+1)T}\right]e^{-at} / \left(1 - e^{aT}\right)$, $NT < t < (N+1)T$.

Using this formula, we see that $y(NT) = c\left[e^{-(N+1)T} - 1\right] / \left(e^{-aT} - 1\right)$, and $y((N+1)T) = ce^{-aT}\left(e^{-(N+1)T} - 1\right) / \left(e^{-aT} - 1\right)$.

As N approaches infinity, both terms approach a constant value, with their difference being $-c$. Thus the amount that is administered at $t = NT$ is absorbed during the ensuing time interval. Thus, the amount of drug in the body is bounded.

5. For every interval, the solution of $y' = -a$ is $y(t) = -at + K$.

For $0 < t < T$, we have the initial condition $y(0) = c$, giving $K = c$ and

$y(t) = -at + c$, $0 < t < T$. Also, $\lim_{t \to T} y(t) = -aT + c$.

For $T < t < 2T$, we have the initial condition $y(T) = c + (-aT + c) = -aT + K$, giving $K = 2c$ and

$y(t) = -at + 2c$, $T < t < 2T$. Also, $\lim_{t \to 2T} y(t) = -2aT + 2c$.

For $2T < t < 3T$, we have the initial condition $y(2T) = c + (-2aT + 2c)$ giving $K = 3c$ and $y(t) = -at + 3c$, $2T < t < 3T$. Also, $\lim_{t \to 3T} y(t) = -3aT + 3c$.

Continuing in this manner gives

$y(t) = -at + (N+1)c$, $NT < t < (N+1)T$. Thus the concentration increases without bound as t increases.

7. $V' = 0$ has the solution $V(t) = C_1$.

$V' = -V/(RC)$ has the solution $V(t) = C_2 e^{-t/(RC)}$.

At time $t = 0$, $V(0) = 0$, so $V(t) = 0$ for $0 < t < T_1$.

At time $t = T_1$, $V(T_1) = E$, so $V(t) = Ee^{-(t-T_1)/(RC)}$ for $T_1 < t < T_1 + T_2$.

At time $t = T_1 + T_2$, $V(0) = 0$, so $V(t) = 0$ for $T_1 + T_2 < t < 2T_1 + T_2$.

At time $t = 2T_1 + T_2$, $V(2T_1 + T_2) = E$, so $V(t) = Ee^{-(t-2T_1-T_2)/(RC)}$ for $2T_1 + T_2 < t < 2T_1 + 2T_2$,

and so on. See Figure 4-1.

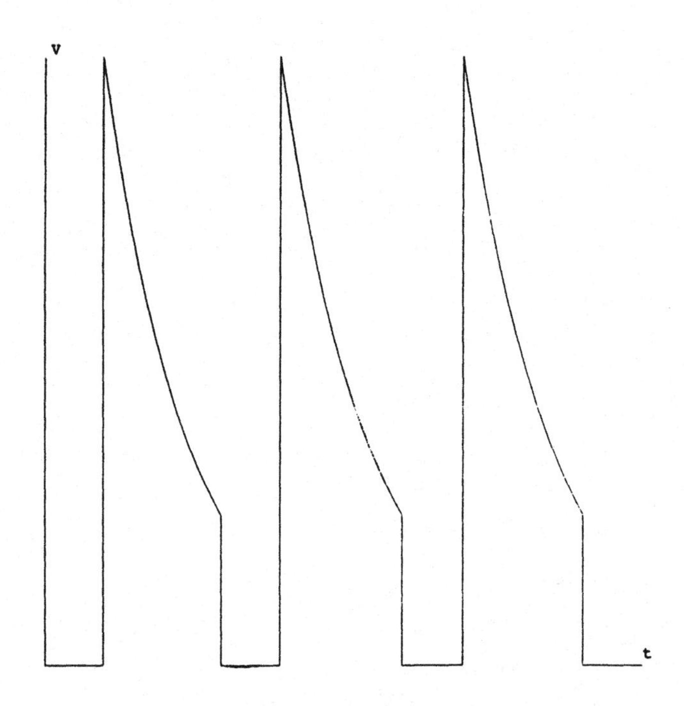

Figure 4-1 Exercise 7.

9. **(a)** If y is differentiable, then $\lim_{x\to k^-} y(x) = \lim_{x\to k^+} [b - y(x)]$, so $y(k) = b/2$.

(b) $y(x) = b$ is an equilibrium solution for $x > k$. Because $y(k) = b/2$, $y' = b - y$ is positive for $x > 0$. Thus, solutions will increase as x increases and approach the equilibrium solution $y(x) = b$. For populations models, b would be the carrying capacity.

(c) For $0 < x < k$, $y(x) = Ce^x$, $y(0) = y_0$, so $C = y_0$, and $y(x) = y_0 e^x$. Now $y(k) = b/2$, so $b/2 = y_0 e^k$, giving $k = \ln(b/(2y_0))$.
For $x > k$, $y'/(b-y) = 1$, so $-\ln(b-y) = x + c$, or $y(x) = b - Ke^{-x}$. $y(k) = b/2$, so $K = (b/2)e^k$, and
$y(x) = b\left(1 - 0.5e^{k-x}\right)$, $x > k$.

(d) From the differential equation,
$y'' = y'$, if $0 < x < k$, and $y'' = -y'$, if $k < x$ Because $y' > 0$ everywhere, this means an inflection point can only occur at $x = k$, $y = b/2$. For given y_0 and b, $k = \ln(b/(2y_0))$, so if $b/(2y_0) < 1$, k will be a negative number, so there will not be an inflection point.

(e) i. If they initially grow at the same rate, then in the logistics equation $a = 1/b$, so from (2.27),
$y(x) = by_0/\left[(b - y_0)e^{-x} + y_0\right]$
ii. If they have the same inflection point, then $y(k) = b/2$ in the logistics equation. This gives $a = -\left[1/(bk)\right]\ln[y_0/(b - y_0)]$.

5. FIRST ORDER LINEAR DIFFERENTIAL EQUATIONS AND MODELS

5.1 Solving Linear Differential Equations

1. (a) $y' + y = x$. $p(x) = 1$, $q(x) = x$.
Because $p(x)$ and $q(x)$ are continuous for all x, we expect the solution to be valid for all x.
Integrating factor is $\mu = \exp \int p(x)\,dx = e^x$.
$e^x y' + e^x y = e^x x$.
$[e^x y]' = e^x x$.
$e^x y = \int x e^x\,dx = (x-1)\,e^x + C$, using integration by parts.
$y(x) = x - 1 + Ce^{-x}$. Valid for all x.

(c) $y' + y = e^x$. $p(x) = 1$, $q(x) = e^x$.
Because $p(x)$ and $q(x)$ are continuous for all x, we expect the solution to be valid for all x.
Integrating factor is $\mu = \exp \int p(x)\,dx = e^x$.
$e^x y' + e^x y = e^x e^x$.
$[e^x y]' = e^{2x}$.
$e^x y = \int e^{2x}\,dx = \frac{1}{2}e^{2x} + C$.
$y(x) = \frac{1}{2}e^x + Ce^{-x}$. Valid for all x.

(e) $y' + 3y = 3x^2 e^{-3x}$. $p(x) = 3$, $q(x) = 3x^2 e^{-3x}$.
Because $p(x)$ and $q(x)$ are continuous for all x, we expect the solution to be valid for all x.
Integrating factor is $\mu = \exp \int p(x)\,dx = e^{3x}$.
$e^{3x} y' + 3e^{3x} y = 3x^2$.
$[e^{3x} y]' = 3x^2$.
$e^{3x} y = \int 3x^2\,dx = x^3 + C$.
$y(x) = x^3 e^{-3x} + Ce^{-3x}$. Valid for all x.

(g) $xy' + 2y = \frac{x}{x^2+2}$.
$y' + \frac{2}{x}y = \frac{1}{x^2+2}$.
$p(x) = \frac{2}{x}$, $q(x) = \frac{1}{x^2+2}$.
Because $p(x)$ is not continuous at $x = 0$, we expect the solution to be valid for all $x \neq 0$.
$\mu(x) = e^{\int 2/x\,dx} = e^{2\ln x} = x^2$.
$x^2 y' + 2xy = \frac{x^2}{x^2+2}$.
$[x^2 y]' = \frac{x^2}{x^2+2}$.
$x^2 y = \int \frac{x^2}{x^2+2}\,dx = \int \left(1 - \frac{2}{x^2+2}\right)dx = x - 2\frac{1}{\sqrt{2}} \arctan \frac{x}{\sqrt{2}} + C$.
$y(x) = \frac{1}{x} - \frac{\sqrt{2}}{x^2} \arctan \frac{x}{\sqrt{2}} + \frac{C}{x^2}$. Valid for all $x \neq 0$.

(i) $y' + \frac{1}{x+1}y = \frac{\cos x}{x+1}$.

$p(x) = \frac{1}{x+1}$, $q(x) = \frac{\cos x}{x+1}$.

Because $p(x)$ and $q(x)$ are not continuous at $x = -1$, we expect the solution to be valid for all $x \neq -1$.

$\mu(x) = e^{\int 1/(x+1)dx} = e^{\ln(x+1)} = x + 1$.

$(x + 1) y' + y = \cos x$.

$[(x + 1) y]' = \cos x$.

$(x + 1) y = \int \cos x \, dx = \sin x + C$.

$y(x) = \frac{\sin x + C}{x+1}$. Valid for all $x \neq -1$.

(k) $y' + 2xy = 2x^3$.

Because $p(x)$ and $q(x)$ are continuous for all x, we expect the solution to be valid for all x.

$p(x) = 2x$, $q(x) = 2x^3$.

$\mu(x) = e^{\int 2x dx} = e^{x^2}$.

$e^{x^2} y' + 2xe^{x^2} y = 2x^3 e^{x^2}$.

$\left[e^{x^2} y\right]' = 2x^3 e^{x^2}$.

$e^{x^2} y = \int 2x^3 e^{x^2} \, dx = x^2 e^{x^2} - e^{x^2} + C$.

$y(x) = x^2 - 1 + Ce^{-x^2}$. Valid for all x.

(m) $y' - \frac{1}{2x}y = 2$.

$p(x) = -\frac{1}{2x}$, $q(x) = 2$.

Because $p(x)$ is not continuous at $x = 0$, we expect the solution to be valid for all $x \neq 0$.

$\mu(x) = e^{\int -1/(2x)dx} = e^{-(\ln x)/2} = |x|^{-1/2}$.

Consider $x > 0$, so that $\mu(x) = x^{-1/2}$.

$x^{-1/2} y' - \frac{1}{2}x^{-3/2} y = 2x^{-1/2}$.

$\left[x^{-1/2} y\right]' = 2x^{-1/2}$.

$x^{-1/2} y = \int 2x^{-1/2} \, dx = 4x^{1/2} + C$.

$y(x) = 4x + Cx^{1/2}$.

In the same way, if $x < 0$, so that $\mu(x) = (-x)^{-1/2}$, then

$y(x) = 4x + C(-x)^{1/2}$.

Combining these we find

$y(x) = 4x + C|x|^{1/2}$. Valid for all $x \neq 0$.

(o) $y' - \frac{2}{x^2}y = \frac{1}{x^2}$.

$p(x) = -\frac{2}{x^2}$, $q(x) = \frac{1}{x^2}$.

Because $p(x)$ and $q(x)$ are not continuous at $x = 0$, we expect the solution to be valid for all $x \neq 0$.

$\mu(x) = e^{\int -2/x^2 \, dx} = e^{2/x}$.

$e^{2/x} y' - \frac{2}{x^2} e^{2/x} y = \frac{1}{x^2} e^{2/x}$.

$\left[e^{2/x} y\right]' = \frac{1}{x^2} e^{2/x}$.

$e^{2/x} y = \int \frac{1}{x^2} e^{2/x} \, dx = -\frac{1}{2} e^{2/x} + C$.

$y(x) = -\frac{1}{2} + Ce^{-2/x}$. Valid for all $x \neq 0$.

(q) $y' + (2 + x^{-1})y = 2e^{-2x}$.

$p(x) = 2 + x^{-1}$, $q(x) = 2e^{-2x}$.

Because $p(x)$ is not continuous at $x = 0$, we expect the solution to be valid for all $x \neq 0$.

$\mu(x) = e^{\int (2+\frac{1}{x})dx} = e^{2x+\ln x} = xe^{2x}$.

$xe^{2x}y' + xe^{2x}(2 + x^{-1})y = 2x$.

$\left[xe^{2x}y\right]' = 2x$.

$xe^{2x}y = \int 2x\,dx = x^2 + C$.

$y(x) = xe^{-2x} + \frac{C}{x}e^{-2x}$. Valid for all $x \neq 0$.

(s) $y' + \frac{4x}{x^2+1}y = 3x$.

$p(x) = \frac{4x}{x^2+1}$, $q(x) = 3x$.

Because $p(x)$ and $q(x)$ are continuous for all x, we expect the solution to be valid for all x.

$\mu(x) = e^{\int 4x/(x^2+1)dx} = \left(x^2 + 1\right)^2$.

$\left(x^2+1\right)^2 y' + 4x\left(x^2+1\right)y = 3x\left(x^2+1\right)^2$.

$\left[\left(x^2+1\right)^2 y\right]' = 3x\left(x^2+1\right)^2$.

$\left(x^2+1\right)^2 y = \int 3x\left(x^2+1\right)^2 dx = \left(x^2+1\right)^3/2 + C$.

$y(x) = \left(x^2+1\right)/2 + C\left(x^2+1\right)^{-2}$. Valid for all x.

3. **(a)** Separable

(c) Separable

(e) Linear

(g) None of these

5. From Exercise 4, $P(t) = P_0/(20\alpha k) + Ce^{kt}$, where $C = P_0\left[1 - 1/(20\alpha k)\right]$. If $P_0 = 1.285$, $\alpha = 4$, $k = 0.0355$, this gives $P(20) = 2.14$ million, compared with the 2.2 for linear emigration.

7. From the slope field, it looks as though all solutions tend to 1 as $x \to \infty$.

Writing the differential equation in the form $y' = 1/\sqrt{x^2+1} - xy/\left(x^2+1\right)$ we see that as $x \to \infty$ both $1/\sqrt{x^2+1}$ and $x/\left(x^2+1\right)$ go to zero. Thus, if $y(x)$ goes to a finite value as $x \to \infty$, then $y' \to 0$, which suggests an asymptote. Furthermore, if $y' = 0$ then $y = \sqrt{1 + 1/x^2}$ and this tends to 1 as $x \to \infty$, which suggests that the asymptote is $y = 1$.

To prove this rigorously, we need the explicit solution of $y' + xy/\left(x^2+1\right) = 1/\sqrt{x^2+1}$.

The integrating factor is $\mu = \exp\left(\int x/(x^2+1)\,dx\right) = \exp\left[\frac{1}{2}\ln\left(x^2+1\right)\right] = \sqrt{x^2+1}$.

Thus,

$\sqrt{x^2+1}y' + xy/\sqrt{x^2+1} = 1$, or

$\left[\sqrt{x^2+1}y\right]' = 1$. Integration gives

$\sqrt{x^2+1}y = x + C$, or

$y(x) = (x + C)/\sqrt{x^2+1}$.

As $x \to \infty$, $y(x) \to 1$ for any C.

5.2 Models That Use Linear Equations

1. **(a)** $y' + 2y = 2e^{-t} \Longrightarrow p(t) = 2$, $q(t) = 2e^{-t} \Longrightarrow$

$\mu(t) = \exp \int 2\,dt = e^{2t}$
$e^{2t}y' + 2e^{2t}y = 2e^t$
$[e^{2t}y]' = 2e^t$
$e^{2t}y = 2e^t + C$
$y(t) = 2e^{-t} + Ce^{-2t}$.

(c) $y' + 2ty = 4t \Longrightarrow p(t) = 2t,\ q(t) = 4t \Longrightarrow$
$\mu(t) = \exp \int 2t\,dt = e^{t^2}$
$e^{t^2}y' + 2te^{t^2}y = 4te^{t^2}$
$[e^{t^2}y]' = 4te^{t^2}$
$e^{t^2}y = 2e^{t^2} + C$
$y(t) = 2 + e^{-t^2}C$.

(e) $y' + (2/t)y = (\sin t)/t^2 \Longrightarrow p(t) = 2/t,\ q(t) = (\sin t)/t^2 \Longrightarrow$
$\mu(t) = \exp \int 2/t\,dt = e^{2\ln t} = t^2$
$t^2y' + 2ty = \sin t$
$[t^2y]' = \sin t$
$t^2y = -\cos t + C$
$y(t) = [-\cos t + C]/t^2$.

(g) $P' + (3/t)P = \ln t/t^2 \Longrightarrow p(t) = 3/t,\ q(t) = \ln t/t^2 \Longrightarrow$
$\mu(t) = \exp \int 3/t\,dt = e^{3\ln t} = t^3$
$t^3P' + 3t^2P = t\ln t$
$[t^3P]' = t\ln t$
$t^3P = \int t\ln t\,dt =$
$\frac{1}{2}t^2\ln t - \int t/2\,dt = \frac{1}{2}t^2\ln t - \frac{1}{4}t^2 + C$ (Using integration by parts)
$P(t) = (2\ln t - 1)/(4t) + C/t^3$.

(i) $P' + \sin tP = 4\cos t\sin t \Longrightarrow p(t) = \sin t,\ q(t) = 4\cos t\sin t \Longrightarrow$
$\mu(t) = \exp(\int \sin t\,dt) = \exp(-\cos t)$
$\exp(-\cos t)P' + \exp(-\cos t)\sin tP = 4\cos t\sin t\exp(-\cos t)$
$[\exp(-\cos t)P]' = 4\cos t\sin t\exp(-\cos t)$
$\exp(-\cos t)P = 4\int \cos t\sin t\exp(-\cos t)\,dt$
$\exp(-\cos t)P = 4\exp(-\cos t)\cos t + 4\int \exp(-\cos t)\sin t\,dt$
$= 4\exp(-\cos t)\cos t + 4\exp(-\cos t) + C$
$P(t) = 4\cos t + 4 + C\exp(\cos t)$.

3. **(a)** $y' - 3y = 6,\ y(0) = 1$.
$\mu(t) = e^{-3t}$ so
$e^{-3t}y' - 3e^{-3t}y = 6e^{-3t}$.
$[e^{-3t}y]' = 6e^{-3t}$
$e^{-3t}y = -2e^{-3t} + C$.
$y(t) = -2 + Ce^{3t}$.
$y(0) = 1 = -2 + C$,
$C = 3$ and
$y(t) = -2 + 3e^{3t}$.

(c) $y' - 7y = 14t,\ y(0) = 0$.
$\mu(t) = e^{-7t}$ so
$e^{-7t}y' - 7e^{-7t}y = 14te^{-7t}$.
$[e^{-7t}y]' = 14te^{-7t}$

$e^{-7t}y = -2te^{-7t} - (2/7)\,e^{-7t} + C,$
$y(t) = -2t - 2/7 + Ce^{7t}.$
$y(0) = 0 = -2/7 + C,$
$C = 2/7$ and
$y\,(t) = -2t - \frac{2}{7} + \frac{2}{7}e^{7t}.$

(e) $y' + 2ty = t,\ y(0) = 2.$
$\mu(t) = e^{t^2}$ so $e^{t^2}y' + 2e^{t^2}y = te^{t^2}.$
$[e^{t^2}y]' = te^{t^2}.$
$e^{t^2}y = e^{t^2}/2 + C,$
$y(t) = 1/2 + Ce^{-t^2}.$
$y(0) = 2 = 1/2 + C,$
$C = 3/2$ and
$y\,(t) = \frac{1}{2} + \frac{3}{2}e^{-t^2}.$

5. Let $x(t)$ be the number of pounds of salt in the container at time t in minutes.

(a) At time $t > 0$ the number of pounds of salt entering the tank per minute is $3 \times 4 = 12$, while the number of pounds of salt leaving per minute is
$[x/\,(200 - t)] \times 5 = 5x/\,(200 - t).$
The differential equation governing this process is
$x' = 12 - 5x/\,(200 - t)$, subject to $x(0) = 0$.
The differential equation is linear with integrating factor
$\mu = \exp\left[\int 5/\,(200 - t)\; dt\right] = (200 - t)^{-5}$, so
$(200 - t)^{-5}\,x' + 5\,(200 - t)^{-6}\,x = 12\,(200 - t)^{-5}$, or
$\left[(200 - t)^{-5}\,x\right]' = 12\,(200 - t)^{-5}.$
Integration gives
$(200 - t)^{-5}\,x = 3\,(200 - t)^{-4} + C$ or
$x(t) = 3\,(200 - t) + C\,(200 - t)^{5}.$
From $x(0) = 0$ we find $C = -3\,(200)^{-4}$ so
$x(t) = 3\,(200 - t) - 3\,(200 - t)^{5}\,/\,(200)^{4}.$
This will be valid until the tank is empty which will occur at $t = 200$ minutes.

(b) The container will contain 2 pounds of salt per gallon when $2 = x/\,(200 - t)$, that is, when $2 = 3 - 3\,(200 - t)^{4}\,/\,(200)^{4}$. Solving for t (in the interval $0 < t < 200$) gives $t = 200 - 200/3^{1/4} \approx 48$ minutes.

7. A possible differential equation is
$T' = -0.2\,[T - 45 - 10\sin(\pi t/12)]$ subject to $T(0) = 70$.
$p(t) = 0.2,\ q(t) = 9 + 2\sin(\pi t/12) \Longrightarrow \mu(t) = \exp(\int 0.2\,dt) = e^{0.2t}$
$e^{0.2t}T' + e^{0.2t}0.2T = e^{0.2t}\,[9 + 2\sin(\pi t/12)]$
$[e^{0.2t}T]' = e^{0.2t}\,[9 + 2\sin(\pi t/12)]$
$e^{0.2t}T = \int e^{0.2t}\,[9 + 2\sin(\pi t/12)]\; dt$
$e^{0.2t}T = 45e^{0.2t} + 2e^{0.2t}\,[0.2\sin(\pi t/12) - \pi/12\cos(\pi t/12)]\,/\,\left(0.04 + \pi^2/144\right) + C$
$T(t) = 45 + 600[(12/5)\sin(\pi t/12) - \pi\cos(\pi t/12)]/\,\left(144 + 25\pi^2\right) + Ce^{-0.2t}$
$T(0) = 45$, so $C = 25 + 600\pi/\,\left(144 + 25\pi^2\right)$
$T(t) = 45{+}600[(12/5)\sin(\pi t/12){-}\pi\cos(\pi t/12)]/[144{+}25\pi^2]{+}\left[25 + 600\pi/\,\left(144 + 25\pi^2\right)\right]e^{-0.2t}$
Graphing this function, we see that T has values less than 40° during the 4 day period. See Figure 5-1. Therefore, the plants would not be safe.

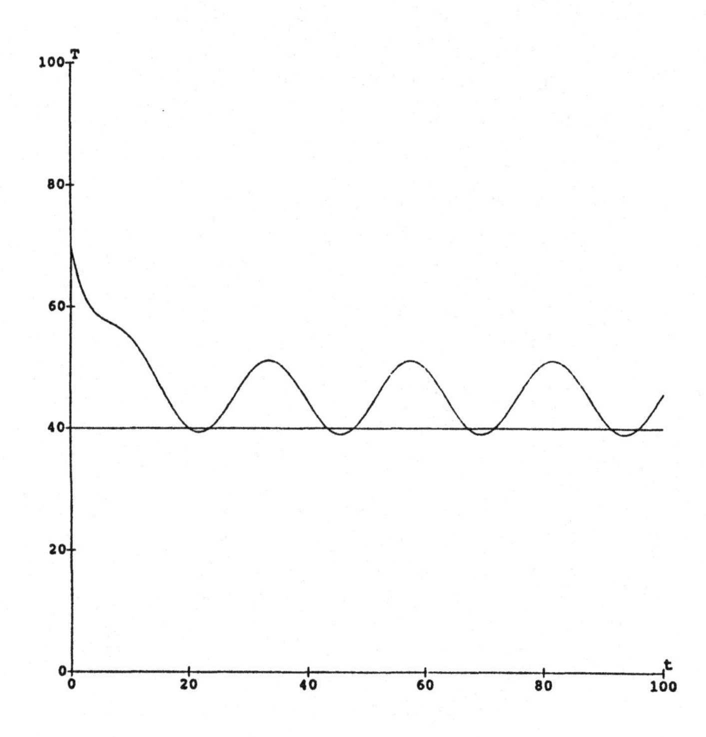

Figure 5-1 Exercise 7.

9. $I' + 60I = 6\sin 2t$, $I(0) = 0$.

$p(t) = 60$, $q(t) = 6\sin 2t \Longrightarrow \mu(t) = \exp(\int 60\,dt) = e^{60t}$.

$e^{60t}I' + 60e^{60t}I = 6e^{60t}\sin 2t$

$\left[e^{60t}I\right]' = 6e^{60t}\sin 2t$

$e^{60t}I = \frac{3}{901}e^{60t}(30\sin 2t - \cos 2t) + C$.

$I(t) = \frac{3}{901}(30\sin 2t - \cos 2t) + Ce^{-60t}$.

$I(0) = 0 \Longrightarrow C = \frac{3}{901}$.

$I(t) = \frac{3}{901}(30\sin 2t - \cos 2t) + \frac{3}{901}e^{-60t}$.

The steady state solution is $\frac{3}{901}(30\sin 2t - \cos 2t)$ and the transient is $\frac{3}{901}e^{-60t}$.

There are no solutions of the form $I(t) = a$, where a is a constant, so there are no equilibrium solutions.

11. $10q' + 100q = 12$, $q(0) = 5$.

$q(t) = 0.12 + Ce^{-10t}$,

$q(0) = 0.12 + C = 5$, so $C = 4.88$ and

$q(t) = 0.12 + 4.88e^{-10t}$.

0.12 is the steady state solution.

Equilibrium solutions would occur if $q(t) = a$ is a solution, where a is a constant. Substituting $q(t) = a$ and $q'(t) = 0$ into $10q' + 100q = 12$ gives $a = 12/100 = 0.12$. Thus, $q(t) = 0.12$ is also the equilibrium solution.

13. **(a)** $y' + y = 2$, $0 \le t < 1$, $y(0) = 0$.

For $0 \le t < 1$,

$y(t) = 2 + Ce^{-t}$; $y(0) = 0$, so $C = -2$ and

$y(t) = 2(1 - e^{-t})$. This gives $\lim_{t\to 1} y(t) = 2(1 - 1/e)$.
For $t \geq 1$, $y' + y = 1$, $y(1) = 2(1 - 1/e)$.
$y(t) = 1 + Ce^{-t}$,
$y(1) = 1 + C/e = 2(1 - 1/e)$, so $C = e - 2$ and
$y(t) = 1 + (e - 2)e^{-t}$.

(c) $y' + y = f(t)$, $y(0) = 6$, $\mu(t) = e^t$ so $e^t y'(t) + e^t y = e^t f(t)$ and
$e^t y(t) = \int e^t f(t)\,dt + C$.
If we write the integral as going from 0 to t we can satisfy the initial condition. This gives
$e^t y(t) = \int_0^t e^u f(u)\,du + y(0)$ or
$y(t) = e^{-t}\left[\int_0^t e^u f(u)\,du + 6\right]$.
The value of this integral depends on whether t is smaller or greater than 10.
For $0 \leq t < 10$, $f(t) = 5$, so
$y(t) = e^{-t}\left[\int_0^t e^u 5\,du + 6\right] = e^{-t}\left[(e^t - 1)\,5 + 6\right] = 5 + e^{-t}$.
For $t \geq 10$,
$y(t) = e^{-t}\left[\int_0^{10} e^u 5\,du + \int_{10}^t e^u 1\,du + 6\right] = e^{-t}\left[\left(e^{10} - 1\right)5 + e^t - e^{10} + 6\right]$.
Thus, $y(t) = e^{-t}\left(4e^{10} + 1\right) + 1$.

15. **(a)** $x' - x/(2t) = V/2$, $p(t) = -1/(2t)$, $q(t) = V/2$. $\mu(t) = t^{-1/2}$, so
$t^{-1/2}x' - t^{-1/2}x/(2t) = t^{-1/2}V/2$
$[t^{-1/2}x]' = t^{-1/2}V/2$.
$t^{-1/2}x = t^{1/2}V + C$ and
$x(t) = tV + C\sqrt{t}$.
If $x(x_0/V) = -x_0 = (x_0/V)V + C\sqrt{x_0/V}$, then $C = -2\sqrt{Vx_0}$ and
$x(t) = tV - 2\sqrt{Vx_0 t}$.
The velocity of the car is given by $x'(t) = V - \sqrt{Vx_0/t}$, showing that its maximum value will be V.

(b) At the traffic light, $x(t) = 0$, so $0 = tV - 2\sqrt{Vx_0 t}$. This gives $\sqrt{t} = 2\sqrt{Vx_0}/V$, or $t = 4x_0/V$. This compares with x_0/V for traveling at the maximum velocity.

17. Let $x(t)$ and $V(t)$ be the height and volume (respectively) of water in the container at time t. Then

$V' = 3 - k\sqrt{x}$, $V(0) = x(0) = 0$, where $V(t) = 25\pi x(t)$.

$25\pi x' = 3 - k\sqrt{x}$, or

$x'/[3 - k\sqrt{x}] = 1/(25\pi)$.

To integrate let $u = k\sqrt{x}$, so $x = u^2/k^2$, $x' = 2uu'/k^2$ and $3 - k\sqrt{x} = 3 - u$. Thus, we have

$[2uu'/k^2]/[3 - u] = 1/(25\pi)$ or

$uu'/[3 - u] = k^2/(50\pi)$. Write this as

$[-1 + 3/(3 - u)]u' = k^2/(50\pi)$ and integrate to obtain

$-u - 3\ln|3 - u| = tk^2/(50\pi) + C$ or

$-k\sqrt{x} - 3\ln|3 - k\sqrt{x}| = tk^2/(50\pi) + C$.

At $t = 0$, $x = 0$, so $C = -\ln 27$. Rearranging gives

$k\sqrt{x} + \ln|3 - k\sqrt{x}|^3 - \ln 27 = -tk^2/(50\pi)$, or

$e^{k\sqrt{x}}\,|3 - k\sqrt{x}|^3/27 = e^{-tk^2/(50\pi)}$.

As $t \to \infty$ the right-hand side goes to zero, so $\lim_{t\to\infty}(3 - k\sqrt{x(t)}) = 0$ or $\lim_{t\to\infty} x(t) = (3/k)^2$. Thus, if $(3/k)^2 < 10$ the container will never be filled.

19. If $x(t)$ is the depth of the water in the container at time t, and $r(t)$ is the radius of the cone at that depth, then by similar triangles $r/x = 5/10$, or $r = x/2$.

The volume of the cone is $V(t) = \pi r^2 x/3 = \pi x^3/12$.

$V' = (\pi/4)\, x^2 dx/dt = 3 - kx$.

This differential equation has the equilibrium solution $x(t) = k/3$.

Thus, if $3/k < 10$, the container will never be filled because for $x > 0$ this differential equation satisfies the conditions of the Existence-Uniqueness Theorem. This means that no other solution can cross this equilibrium solution.

21. Let $h(t)$ and $V(t)$ be the height and volume (respectively) of water in the container at time t. Let $x\,(t)$ be the amount of salt in the container at time t.

$V = \pi r^2 h = 64h$, so

$V' = 64h' = 4 - 0.2h$.

Notice that the equilibrium solution $h(t) = 20$ is greater than the height of the container, so it is of no concern.

$64h' = 4 - 0.2h$ can be written

$320h' = 20 - h$ giving

$h'/\,(20 - h) = 1/320$,

$-\ln(20 - h) = t/320 + c$

$h = 20 - Ke^{-t/320}$.

$h(0) = 5 = 20 - K$, so

$K = 15$ and

$h(t) = 20 - 15e^{-t/320}$.

$x' = 12 - 0.2h(t)x/V(t)$

$= 12 - 0.2h(t)x/[64h(t)]$

$= 12 - x/320$

$= (3840 - x)\,/320$, so

$x'/\,(3840 - x) = 1/320$,

$-\ln|3840 - x| = t/320 + c$

$x(t) = 3840 + Ke^{-t/320}$.

$x(0) = 0$, so $K = -3840$ and

$x(t) = 3840\left(1 - e^{-t/320}\right)$.

$x(t)/V(t) = x(t)/[64h(t)] = (1/2)3$ when

$3840\left(1 - e^{-t/320}\right) / \left[64\left(20 - 15e^{-t/320}\right)\right] = 1.5$.

A graphical or numerical solution gives $t \approx 71.4$ minutes.

$h\,(71.4) \approx 8$ feet. Thus, the container has not overflowed.

23. The equilibrium solution is $x_\infty = (Cr - R)/k$, so solving for the reaction rate gives $k = (Cr - R)/x_\infty$.

25. Letting $t = x_i u$ gives $dt = x_i du$ and

$$\frac{x_i}{2x_i^2 - 1} = e^{-x_i^2} \int_0^{x_i} e^{t^2}\, dt$$

becomes

$$\frac{x_i}{2x_i^2 - 1} = e^{-x_i^2} \int_0^1 e^{x_i^2 u^2} x_i \, du,$$

or

$$1 = e^{-x_i^2} \int_0^1 \left(2x_i^2 - 1\right) e^{x_i^2 \left(u^2 - 1\right)} \, du.$$

Experimenting gives $x_i \approx 1.502$.

27. $y' + 2xy = 1$

$y(x) = \sum_{k=0}^{\infty} c_k x^k$

$y'(x) = \sum_{k=0}^{\infty} c_k k x^{k-1}$

Notice that $y(0) = 0$ implies $c_0 = 0$.

$\sum_{k=0}^{\infty} c_k k x^{k-1} + 2x \sum_{k=0}^{\infty} c_k x^k = 1$

$c_1 + \sum_{k=2}^{\infty} c_k k x^{k-1} + \sum_{k=2}^{\infty} 2c_{k-2} x^{k-1} = 1.$

$c_1 + \sum_{k=2}^{\infty} \left(c_k k + 2c_{k-2}\right) x^{k-1} = 1.$

$c_1 = 1,\ kc_k + 2c_{k-2} = 0,\ k = 2, 3, 4 \cdots.$

$c_k = -\frac{2}{k} c_{k-2},\ k = 2, 3, 4$

$c_2 = -c_0 = 0,$

$c_3 = -2c_1/3 = -2/3,$

$c_4 = -2c_2/4 = (-1)^2 c_0/2! = 0,$

$c_5 = -2c_3/5 = (-2)^2 c_1/(5 \cdot 3) = (-2)^2/(5 \cdot 3),$

$c_6 = -2c_4/6 = (-1)^3 c_0/3! = 0,$

$c_7 = -2c_3/5 = (-2)^3 c_1/(7 \cdot 5 \cdot 3) = (-2)^3/(7 \cdot 5 \cdot 3), \cdots$

$y(x) = c_0 + c_1 x + c_2 x^2 + c_3 x^3 + c_4 x^4 + c_5 x^5 + c_6 x^6 + c_7 x^7 + \cdots$

$= x - 2x^3/3 + (-2)^2 x^5/(5 \cdot 3) + (-2)^3 x^7/(7 \cdot 5 \cdot 3) + \cdots$

$= \sum_{k=0}^{\infty} \frac{(-1)^k 2^k}{1 \cdot 3 \cdots (2k+1)} x^{2k+1}.$

Plotting the terms up to x^7 gives excellent agreement for $0 \leq x \leq 1$.

5.3 Models That Use Bernoulli's Equation

1. **(a)** $yy' + xy^2 = x,$

$u = y^2,$

$u' = 2yy',$

$\frac{1}{2}u' + xu = x$, or

$u' + 2xu = 2x$

$\mu(x) = e^{x^2}$

$e^{x^2} u' + e^{x^2} 2xu = e^{x^2} 2x$

$[e^{x^2} u]' = e^{x^2} 2x$

$e^{x^2} u = e^{x^2} + C,$

$u = 1 + Ce^{-x^2}.$

$y^2 = 1 + Ce^{-x^2}.$

(c) $xy' - (1+x)y = y^2,$

$xy^{-2}y' - (1+x)y^{-1} = 1.$

$u = y^{-1},$

$u' = -y^{-2}y'.$

$-xu' - (1+x)u = 1,$

$u' + (1/x + 1)u = -1/x,$

$\mu(x) = \exp\left[\int (1/x + 1)\, dx\right] = \exp(\ln x + x) = xe^x.$

$xe^x u' + e^x(1+x)u = -e^x$

$[xe^x u]' = -e^x$

$xe^x u = -e^x + C,$

$u = (-e^x + C)/(xe^x).$

$y(x) = 1/u = xe^x/(-e^x + C)$

(e) $y' + 2y/x = -x^9 y^5,$

$y^{-5}y' + 2y^{-4}/x = -x^9$

$u = y^{-4},$

$u' = -4y^{-5}y',$

$-(1/4)\, u' + 2u/x = -x^9,$

$u' - 8u/x = 4x^9,$

$\mu(x) = \exp(\int -8/x)\, dx = x^{-8}$

$x^{-8}u' - 8x^{-7}u = 4x$

$[x^{-8}u]' = 4x$

$x^{-8}u = 2x^2 + C$

$u = 2x^{10} + Cx^8$

$y^4 = 1/\left(2x^{10} + Cx^8\right)$

(g) $y' + \sqrt{x}y = (2/3)\sqrt{x/y},$

$y^{1/2}y' + \sqrt{x}y^{3/2} = (2/3)\sqrt{x}.$

$u = y^{3/2},$

$u' = (3/2)y^{1/2}y',$

$(2/3)\, u' + \sqrt{x}u = (2/3)\sqrt{x},$

$u' + (3/2)\sqrt{x}u = \sqrt{x},$

$\mu(x) = \exp\left(\int (3/2)\sqrt{x}\, dx\right) = e^{x^{3/2}}$

$e^{x^{3/2}}u' + e^{x^{3/2}}(3/2)\sqrt{x}u = \sqrt{x}e^{x^{3/2}}$

$\left[e^{x^{3/2}}u\right]' = \sqrt{x}e^{x^{3/2}}$

$e^{x^{3/2}}u = (2/3)\, e^{x^{3/2}} + C$

$u = 2/3 + Ce^{-x^{3/2}} = y^{3/2}$

$y(x) = \left(2/3 + Ce^{-x^{3/2}}\right)^{2/3}$

(i) $y' - y = -xe^{-2x}y^3$

$y^{-3}y' - y^{-2} = -xe^{-2x}$

$u = y^{-2},$

$u' = -2y^{-3}y'.$

$-(1/2)\, u' - u = -xe^{-2x},$

$u' + 2u = 2xe^{-2x},$

$\mu(x) = \exp(\int 2\, dx) = e^{2x}.$

$e^{2x}u' + e^{2x}2u = 2x$

$$[e^{2x}u]' = 2x$$
$$e^{2x}u = x^2 + C,$$
$$u = (x^2 + C)/e^{2x} = 1/y^2.$$
$$y^2 = 1/u = e^{2x}/\left(x^2 + C\right)$$

3. The equation $w' = Hw^m - kw$ can be written

$$\frac{1}{k}\left(\frac{Hw^{m-2}}{Hw^{m-1} - k} - \frac{1}{w}\right) w' = 1,$$

which, when integrated yields

$$\frac{1}{m-1} \ln\left|Hw^{m-1} - k\right| - \ln|w| = kt + c,$$

or

$$\ln\left|\frac{Hw^{m-1} - k}{w^{m-1}}\right| = (m-1)(kt + c),$$

which can be solved for $w(t)$.

For $H = 1$, $m = 2/3$, $k = 0.6$, we have

$$\frac{1}{0.6}\left(\frac{w^{-4/3}}{w^{-1/3} - 0.6} - \frac{1}{w}\right) w' = 1,$$

so

$-3\ln\left|w^{-1/3} - 0.6\right| - \ln|w| = 0.6t + c$, or

$\ln\left|w^{-1/3} - 0.6\right| + (1/3)\ln|w| = -0.2t - c/3$

$(w^{-1/3} - 0.6)w^{1/3} = Ce^{-0.2t}$

$1 - 0.6w^{1/3} = Ce^{-0.2t}$

$w(t) = \left[\left(1 - Ce^{-0.2t}\right)/0.6\right]^3$, which is (5.67).

5. **(a)** The equilibrium solutions are $y(x) = 0$ (stable) and $y(x) = 1/m$ (unstable).

(b) The equilibrium solutions are $y(x) = 0$ (stable), $y(x) = 1/m$ (unstable), and $y(x) = -1/m$ (unstable).

(c) The equilibrium solutions are $y(x) = 0$ (stable), $y(x) = 1/m$ (unstable), and $y(x) = -1/m$ (unstable).

(d) $y' = y[(my)^{n-1} - 1]$, so near $y = 0$, $(my)^{n-1}$ has values close to 0 for all positive values of m and n except $n = 1$. Thus, $[(my)^{n-1} - 1]$ is negative for y near 0 and the equilibrium solution $y(x) = 0$ will always be stable. If $n = 1$, the differential equation becomes $y' = 0$ which has $y(x) = 0$ as a stable solution also.

7. $P' - k(t)P = -\left[k(t)/b(t)\right]P^2$, or

$P^{-2}P' - k(t)P^{-1} = -k(t)/b(t)$.

$u = P^{-1}$,

$u' = -P^{-2}P'$, so

$-u' - k(t)u = -k(t)/b(t)$.

$\mu(t) = \exp(\int k(t)\, dt)$

$\mu(t)u' + \mu(t)k(t)u = \mu(t)k(t)/b(t)$

$[\mu(t)u]' = \mu(t)k(t)/b(t)$

$\mu(t)u = \int \mu(t)k(t)/b(t)\, dt + C$

$u = [1/\mu(t)][\int \mu(t)k(t)/b(t)\, dt + C]$

$P(t) = \mu(t)/[\int \mu(t)k(t)/b(t)dt + C]$.

A possible application is a situation where both the growth rate, k, and the carrying capacity, b, depend on the time of the year.

9. **(a)** Separable

(c) None of these

(e) Separable

(g) Bernoulli, homogeneous coefficients

(i) Linear

6. INTERPLAY BETWEEN FIRST ORDER SYSTEMS AND SECOND ORDER DIFFERENTIAL EQUATIONS

6.1 Simple Models

1. The ratio of their emotional ranges is y/x. In general

$x(t) = C_1 \cos\sqrt{2}t + C_2 \sin\sqrt{2}t$ and $y(t) = -\sqrt{2}C_1 \sin\sqrt{2}t + \sqrt{2}C_2 \cos\sqrt{2}t$.

If $x(0) = 0$ then $C_1 = 0$, and $y/x = \left(\sqrt{2}C_2 \cos\sqrt{2}t\right) / \left(C_2 \cos\sqrt{2}t\right) = \sqrt{2}$, which is independent of C_2. But $y(0) = C_2$, so y/x is independent of Chad's initial emotion.

3. Because $x' = -x + y$, $x(t)$ will increase with time if $y > x$ and decrease if $y < x$.

Because $y' = -x - y$, $y(t)$ will increase with time if $y < -x$ and decrease if $y > -x$.

If we concentrate on the region above the lines $y = x$ and $y = -x$, then in this region x will be increasing and y will be decreasing. Thus, arrows will be pointing down to the right. Similar arguments can be used on the remaining three regions, and the arrows overall spiral clockwise. See Figure 6-1.

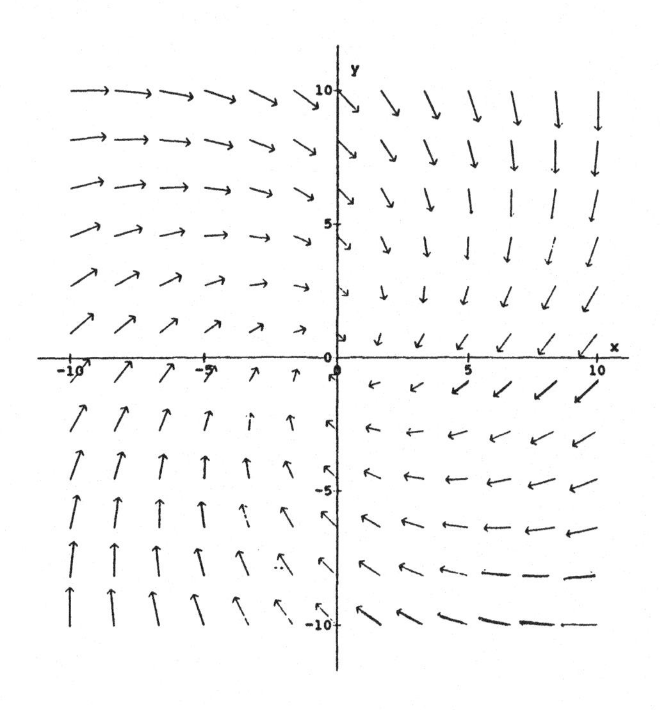

Figure 6-1 Exercise 3.

5. Using $y = zx$ in $y' = -2x/(y-0.4x)$ gives $z'x + z = -2/(z-0.4)$, or

$z' = -\left(5z^2 - 2z + 10\right)/(5z-2)$.

There are no equilibrium solutions.

Nonequilibrium solutions are given by

$\int (5z-2)/\left(5z^2-2z+10\right)\, dz = -\int dx/x = -\ln|x|$.

By writing the integral on the left-hand side as

$\int (5z-1)/\left(5z^2-2z+10\right)\, dz - \int 1/\left(5z^2-2z+10\right)\, dz$,

and making the substitution $u = 5z^2 - 2z + 10$ in the first integral, and $z = \frac{1}{5} + \frac{7}{5}\tan v$ in the second, we find

$\int (5z-2)/\left(5z^2-2z+10\right)\, dz = \frac{1}{2}\ln\left(5z^2-2z+10\right) - \frac{1}{7}\arctan\left(\frac{5}{7}z - \frac{1}{7}\right)$,

so the equation for the orbits is

$\frac{1}{2}\ln\left[5\left(\frac{y}{x}\right)^2 - 2\frac{y}{x} + 10\right] - \frac{1}{7}\arctan\left(\frac{5y}{7x} - \frac{1}{7}\right) = -\ln|x| + C.$

6.2 How First Order Systems and Second Order Equations Are Related

1. (a) Linear, constant coefficients, nonhomogeneous, nonautonomous.

   (c) Linear, homogeneous, nonautonomous.

   (e) Nonlinear, autonomous.

3. If $x'' = R(t, x, x')$ is autonomous, then $x'' = R(x, x')$ and the corresponding system is $x' = y$, $y' = R(x, y)$, which is autonomous.

   If $x'' = R(t, x, x')$ is linear, then $x'' = -q(t)x - p(t)x' + f(t)$ and the corresponding system is $x' = y$, $y' = -q(t)x - p(t)y + f(t)$, which is linear.

   If $x'' = R(t, x, x')$ is homogeneous, then $x'' = -q(t)x - p(t)x'$ and the corresponding system is $x' = y$, $y' = -q(t)x - p(t)y$, which is homogeneous.

   If $x'' = R(t, x, x')$ has constant coefficients, then $x'' = -ax - bx' + f(t)$, where a and b are constants, and the corresponding system is $x' = y$, $y' = -ax - by + f(t)$, which has constant coefficients.

5. Substituting $x' = y$ into $x'' = R(t, x')$ gives $y' = R(t, y)$.

7. (a) Substituting $x' = y$ into $x'' + a^2\left(1-x^2\right)x' + x = 0$ gives the system $x' = y$, $y' = -a^2\left(1-x^2\right)y - x$.

   (b) Substituting $x' = y - a^2\left(x - \frac{1}{3}x^3\right)$ and $x'' = y' - a^2\left(1-x^2\right)x'$ into $x'' + a^2\left(1-x^2\right)x' + x = 0$ gives the system $x' = y - a^2\left(x - \frac{1}{3}x^3\right)$, $y' = -x$.

9. Substituting $x' = y$ into $x'' = g - \frac{k}{m}\left(x'\right)^2$ gives the system $x' = y$, $y' = g - \frac{k}{m}y^2$. Solving the second equation for $y(t)$ and substituting the result in the first, is exactly the way we solved this equation in Section 4.4.

11. We are given $a_2(t)x_1'' + a_1(t)x_1' + a_0(t)x_1 = 0$, and $a_2(t)x_2'' + a_1(t)x_2' + a_0(t)x_2 = 0$, so that

$$a_2(t)\left(C_1x_1 + C_2x_2\right)'' + a_1(t)\left(C_1x_1 + C_2x_2\right)' + a_0(t)\left(C_1x_1 + C_2x_2\right)$$
$$= a_2(t)\left(C_1x_1\right)'' + a_2(t)\left(C_2x_2\right)'' + a_1(t)\left(C_1x_1\right)' + a_1(t)\left(C_2x_2\right)' + a_0(t)\left(C_1x_1\right) + a_0(t)\left(C_2x_2\right)$$
$$= a_2(t)C_1x_1'' + a_2(t)C_2x_2'' + a_1(t)C_1x_1' + a_1(t)C_2x_2' + a_0(t)C_1x_1 + a_0(t)C_2x_2$$

$= C_1\left(a_2(t)x_1'' + a_1(t)x_1' + a_0(t)x_1\right) + C_2\left(a_2(t)x_2'' + a_1(t)x_2' + a_0(t)x_2\right)$

$= 0.$

13. Differentiating $x' = a(t)\,x + b(t)\,y + f(t)$ with respect to t gives

$x'' = a'x + ax' + b'y + by' + f'.$

If $b(t) = 0$, this is a second order equation for $x(t)$.

If $b(t) \neq 0$, then we can use $y' = cx + dy + g$ to find

$x'' = a'x + ax' + b'y + b(cx + dy + g) + f'$, and $y = (x' - ax - f)/b$, to find

$x'' = a'x + ax' + b'(x' - ax - f)/b + b\left[cx + d(x' - ax - f)/b + g\right] + f'$,

which can be written

$x'' - (a + b' + bd)\,x' - (a' - ab' + bc - abd)\,x = -b'f/b - bdf/b + bg + f'.$

15. **(a)** From $x' + y' + y = 0$ we find $x' = -y' - y$, so $3x' + 2y' + x = 7$ can be written $-y' - 3y + x = 7$. Differentiating this with respect to t gives $-y'' - 3y' + x' = 0$, and using $x' = -y' - y$ again gives

$y'' + 4y' + y = 0.$

(c) From $2x' - 2y' - y = 8$ we find $x' = y' + \frac{1}{2}y + 4$, so $x' - 3y' - 4x = e^{2t}$ can be written $-2y' + \frac{1}{2}y + 4 - 4x = e^{2t}$. Differentiating this with respect to t gives $-2y'' + \frac{1}{2}y' - 4x' = 2e^{2t}$, and using $x' = y' + \frac{1}{2}y + 4$ again, gives

$-2y'' - \frac{7}{2}y' - 2y - 16 = 2e^{2t}.$

(e) From $2x' - y' - y = -2\sin t$ we find $x' = \frac{1}{2}y' + \frac{1}{2}y - \sin t$, so $tx' + x + y' = \cos t$ can be written $t\left(\frac{1}{2}y' + \frac{1}{2}y - \sin t\right) + x + y' = \cos t$. Differentiating this with respect to t gives $t\left(\frac{1}{2}y'' + \frac{1}{2}y' - \cos t\right) + \frac{1}{2}y' + \frac{1}{2}y - \sin t + x' + y'' = -\sin t$, and using $x' = \frac{1}{2}y' + \frac{1}{2}y - \sin t$ again gives

$\left(\frac{1}{2}t + 1\right)y'' + \left(\frac{1}{2}t + 1\right)y' + y = \sin t + t\cos t.$

17. The system $\frac{dx}{dt} = -P(x,y)$, $\frac{dy}{dt} = -Q(x,y)$, can be obtained from the system $\frac{dx}{dt} = P(x,y)$, $\frac{dy}{dt} = Q(x,y)$ by replacing t by $-t$. This has the effect of reversing the direction of travel, and the arrows in the direction field. See Figure 6-2.

19. The uniqueness theorem requires two initial conditions, namely $x(0) = 0$ and $x'(0) = x_0^*$. The trivial solution corresponds to $x_0^* = 0$ whereas $x(t) = \sin t$ corresponds to $x_0^* = 1$.

6.3 Second Order Linear Differential Equations with Constant Coefficients

1. **(a)** Characteristic equation $r^2 - r - 6 = 0$ has distinct real roots $r_1 = -2$, $r_2 = 3$.

$x(t) = C_1e^{-2t} + C_2e^{3t}.$

(c) Characteristic equation $r^2 - 2r + 1 = 0$ has a double root $r_1 = r_2 = 1$.

$x(t) = C_1e^{t} + C_2te^{t}.$

(e) Characteristic equation $r^2 + 2r - 8 = 0$ has distinct real roots $r_1 = -4$, $r_2 = 2$.

$x(t) = C_1e^{-4t} + C_2e^{2t}.$

(g) Characteristic equation $r^2 - r - 12 = 0$ has distinct real roots $r_1 = -3$, $r_2 = 4$.

$x(t) = C_1e^{-3t} + C_2e^{4t}.$

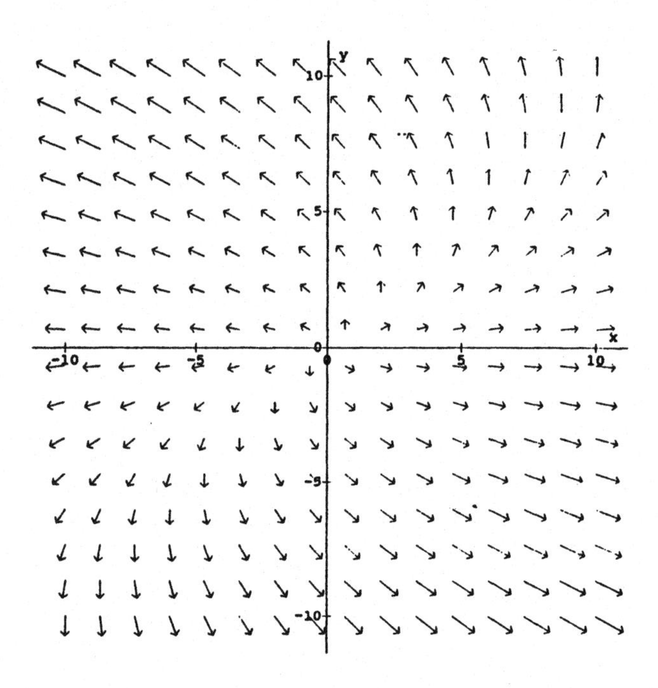

Figure 6-2 Exercise 17.

(i) Characteristic equation $r^2 - 5r + 6 = 0$ has distinct real roots $r_1 = 3$, $r_2 = 2$.
$x(t) = C_1e^{2t} + C_2e^{3t}$.

(k) Characteristic equation $r^2 + 16 = 0$ has complex conjugate roots $\pm 4i$.
$x(t) = C_1 \cos 4t + C_2 \sin 4t$.

(m) Characteristic equation $6r^2+r-1 = 0$ has distinct real roots $r_1 = -1/2$, $r_2 = 1/3$.
$x(t) = C_1e^{-t/2} + C_2e^{t/3}$.

(o) Characteristic equation $r^2 + 2r + 4 = 0$ has complex conjugate roots $-1 \pm \sqrt{3}i$.
$x(t) = e^{-t}\left(C_1 \cos \sqrt{3}t + C_2 \sin \sqrt{3}t\right)$.

3. **(a)** Characteristic equation $r^2 + r - 2 = 0$ has roots $r_1 = 1$, $r_2 = -2$ and solution
$x(t) = C_1e^t + C_2e^{-2t}$.
$x(0) = 0 \Longrightarrow 0 = C_1 + C_2$.
$x'(t) = C_1e^t - 2C_2e^{-2t}$.
$x'(0) = 3 \Longrightarrow 3 = C_1 - 2C_2$.
Solving for C_1 and C_2 gives $C_1 = 1$, $C_2 = -1$ so
$x(t) = e^t - e^{-2t}$.

(c) Characteristic equation $r^2 + 6r + 9 = 0$ has a double root $r = -3$ and solution
$x(t) = C_1e^{-3t} + C_2te^{-3t}$.
$x(0) = 2 \Longrightarrow 2 = C_1$.
$x'(t) = -3C_1e^{-3t} + C_2\left(e^{-3t} - 3te^{-3t}\right)$.
$x'(0) = 0 \Longrightarrow 0 = -3C_1 + C_2$.
Solving for C_1 and C_2 gives $C_1 = 2$, $C_2 = 6$ so
$x(t) = 2e^{-3t} + 6te^{-3t}$.

(e) Characteristic equation $r^2 + 3r = 0$ has roots $r_1 = 0$, $r_2 = -3$ and solution
$x(t) = C_1 + C_2 e^{-3t}$.
$x(0) = 4 \Longrightarrow 4 = C_1 + C_2$.
$x'(t) = -3C_2 e^{-3t}$.
$x'(0) = 3 \Longrightarrow 3 = -3C_2$.
Solving for C_1 and C_2 gives $C_1 = 5$, $C_2 = -1$ so
$x(t) = 5 - e^{-3t}$.

(g) Characteristic equation $r^2 + 10r + 100 = 0$ has roots $r_{1,2} = -5 \pm i5\sqrt{3}$ and solution
$x(t) = C_1 e^{-5t} \cos 5\sqrt{3}t + C_2 e^{-5t} \sin 5\sqrt{3}t$.
$x(0) = 15 \Longrightarrow 15 = C_1$.
$x'(t) = C_1 e^{-5t}\left(-5\cos 5\sqrt{3}t - 5\sqrt{3}\sin 5\sqrt{3}t\right) + C_2 e^{-5t}\left(-5\sin 5\sqrt{3}t + 5\sqrt{3}\cos 5\sqrt{3}t\right)$.
$x'(0) = 4 \Longrightarrow 4 = -5C_1 + 5\sqrt{3}C_2$.
Solving for C_1 and C_2 gives $C_1 = 15$, $C_2 = 79\sqrt{3}/15$ so
$x(t) = 15e^{-5t}\cos 5\sqrt{3}t + \frac{79}{15}\sqrt{3}e^{-5t}\sin 5\sqrt{3}t$.

5. Characteristic equation $r^2 + 0.4r + 2 = 0$ has complex roots $r_{1,2} = -0.2 \pm 1.4i$, and solution

$x(t) = C_1 e^{-0.2t}\cos(1.4t) + C_2 e^{-0.2t}\sin(1.4t)$.

7. $e^{\alpha t}(C_1 \cos\beta t + C_2 \sin\beta t)$
$= \frac{1}{2}e^{\alpha t}\left[C_1\left(e^{i\beta t} + e^{-i\beta t}\right) + \frac{1}{i}C_2\left(e^{i\beta t} - e^{-i\beta t}\right)\right]$
$= \frac{1}{2}e^{\alpha t}\left[e^{i\beta t}(C_1 - iC_2) + e^{-i\beta t}(C_1 + iC_2)\right]$
$= Ke^{\alpha t + i\beta t} + \tilde{K}e^{\alpha t - i\beta t}$.

9. (a) $r^2 + 9 = 0$, $r = \pm 3i$, so $x(t) = A\cos(3t + \phi)$.
$x(0) = A\cos\phi = -2$, and $x'(0) = -3A\sin\phi = -6$, gives $A\cos\phi = -2$ and $A\sin\phi = 2 \Longrightarrow A^2(\cos^2\phi + \sin^2\phi) = 8$ or $A = \sqrt{8}$.
Division gives $\tan\phi = -1$, so $\phi = 3\pi/4$ and
$x(t) = \sqrt{8}\cos(3t + 3\pi/4)$.
Amplitude $= \sqrt{8}$, Period $= 2\pi/3$, Phase angle $= 3\pi/4$.

11. The equation $x'' + ax' + bx = 0$ only has three possible solutions, namely,

(i) $x(t) = e^{\alpha t}(C_1\cos\beta t + C_2\sin\beta t)$, which can cross the t-axis an infinite number of times.

(ii) $x(t) = (C_1 + C_2 t)e^{rt}$, which can cross the t-axis either once or no times, depending on whether $C_1 + C_2 t = 0$ has one or zero roots.

(iii) $x(t) = C_1 e^{r_1 t} + C_2 e^{r_2 t}$, which can cross the t-axis either once or no times, depending on whether $C_1 e^{r_1 t} + C_2 e^{r_2 t}$ has one or zero roots.

Thus, the solution may cross the t-axis, exactly no times, once, or an infinite number of times.

13. We want to choose the constant c so that the solution of $x' = y - cx$, $y' = -2x$, $x(0) = 0$, $y(0) = 2.5$, does not oscillate, and, furthermore, both x and y are positive for $t > 0$. This system gives rise to $x'' = y' - cx' = -2x - cx'$, or $x'' + cx' + 2x = 0$, with characteristic equation $r^2 + cr + 2 = 0$. This will have nonoscillating solutions if $c^2 - 8 \geq 0$, with $c = 2\sqrt{2}$, being the smallest value of c for this to happen. In this case, we will have repeated roots of $r = -\sqrt{2}$, and solution $x(t) = (C_1 + C_2 t)e^{rt}$. The initial condition $x(0) = 0$ requires that $C_1 = 0$, so, in this case, $x(t) = C_2 te^{rt}$, which is never zero for $t > 0$. Substituting this into $y = x' + 2\sqrt{2}x$ (where $r = -\sqrt{2}$) gives $y(t) = C_2\left(e^{rt} + \sqrt{2}te^{rt}\right)$ which is never zero for $t > 0$. Thus, $c = 2\sqrt{2}$ is the smallest value of c that satisfies the conditions. Notice it is critical damping!

6.4 Modeling Physical Situations

1. **(a)** $x'' + \frac{1}{8}x' + x = 0 \Longrightarrow r^2 + \frac{1}{8}r + 1 = 0 \Longrightarrow r = \left(-\frac{1}{8} \pm \sqrt{\frac{1}{64} - 4}\right)/2 \Longrightarrow$
$r = \left(-1 \pm \sqrt{255}i\right)/16 = \alpha \pm i\beta \Longrightarrow$
$x(t) = C_1 e^{\alpha t} \cos \beta t + C_2 e^{\alpha t} \sin \beta t.$
$x(0) = 0 = C_1$, $x'(0) = C_2 \beta = \frac{1}{2}$, so $C_2 = 1/(2\beta)$ and
$x(t) = \left(8/\sqrt{255}\right) e^{-t/16} \sin(\sqrt{255}t/16).$

(c) $64r^2 + 16r + 17 = (8r + 1)^2 + 16 = 0$, so $r = (-1 \pm 4i)/8$ and
$x(t) = C_1 e^{-t/8} \cos t/2 + C_2 e^{-t/8} \sin t/2.$
$x(0) = 1 = C_1$, $x'(0) = -C_1/8 + C_2/2 = 0 \Longrightarrow C_2 = 1/4$ and
$x(t) = e^{-t/8} \cos t/2 + (1/4)e^{-t/8} \sin t/2.$

(e) $x(t) = 0$. This follows immediately from the uniqueness theorem.

3. The characteristic equation is $4r^2 + br + 9 = 0$ with solutions $r = -b/8 \pm (1/8)\sqrt{b^2 - 144}$. Motion will be

(a) overdamped if $b > 12$

(b) underdamped if $0 < b < 12$

(c) critically damped if $b = 12$.

5. The characteristic equation is $mr^2 + br + k = 0$ with solutions $r = \left(-b \pm \sqrt{b^2 - 4mk}\right)/(2m)$.

If $b^2 - 4mk < 0$ then

$x(t) = e^{-bt/(2m)} \left\{C_1 \cos\left[\sqrt{4mk - b^2}t/(2m)\right] + C_2 \sin\left[\sqrt{4mk - b^2}t/(2m)\right]\right\}$

which approaches 0 as $t \to \infty$, because $-b/(2m) < 0$.

If $b^2 - 4mk = 0$ then

$x(t) = e^{-bt/(2m)} (C_1 + C_2 t)$

which approaches 0 as $t \to \infty$ because $-b/(2m) < 0$.

If $b^2 - 4mk > 0$ then

$x(t) = C_1 e^{r_1 t} + C_2 e^{r_2 t}$, where $r_{1,2} = \left(-b \pm \sqrt{b^2 - 4mk}\right)/(2m)$

which approaches 0 as $t \to \infty$ because $r_1 < 0$ and $r_2 < 0$.

7. $x(t) = A\cos(\omega t + \phi)$, $x'(t) = -\omega A \sin(\omega t + \phi)$, $x''(t) = -\omega^2 A \cos(\omega t + \phi)$, so
$x''(t) + \omega^2 x = -\omega^2 A \cos(\omega t + \phi) + \omega^2 A \cos(\omega t + \phi) = 0$. Here A and ϕ are the two arbitrary constants, and these are the ones that are specified by the two initial conditions.

9. If $E = \frac{1}{2}(x')^2 + \frac{1}{2}kx^2$ then $E' = x'x'' + kxx' = x'(-bx' - kx) + kxx' = -b(x')^2$. Because $E' < 0$ then E is a decreasing function of time.

11. The straight line of best fit through the data points in Figure 6.16 of the text has slope 5.74 and intercept 0.00, so $x = 5.74m$.

The straight line of best fit through the data points in Figure 6.17 of the text has slope 2.97 and intercept 0.00, so $x = 2.97m$.

Both of these satisfy Hooke's law.

13. The motion of the pendulum is governed by $x'' + \frac{g}{h}x = 0$, where h is the length of the pendulum. The solution of this equation can be written $x(t) = A\cos(\lambda t + \phi)$, where $\lambda = \sqrt{\frac{g}{h}}$. Thus, $x(t)$ attains a maximum when $\lambda t + \phi = 2n\pi$, $n = 0, \pm 1, \pm 2, \cdots$. The time between successive maximums is $2\pi/\lambda = 2\pi\sqrt{\frac{h}{g}}$. If I want to shorten this time, I should make h smaller, that is, shorten the pendulum.

15. The characteristic equation is $mr^2+br+k = 0$ with solutions $r = \left(-b \pm \sqrt{b^2 - 4mk}\right)/(2m)$. For the underdamped case, $b^2 - 4mk < 0$, so $r = \alpha \pm i\beta$, where $\alpha = -b/(2m)$ and $\beta = \sqrt{4mk - b^2}/(2m)$, and the solution can be written $x(t) = Ae^{\alpha t}\cos(\beta t + \phi)$. From this, we have $x'(t) = A\alpha e^{\alpha t}\cos(\beta t + \phi) - \beta A e^{\alpha t}\sin(\beta t + \phi)$. The initial condition $x(0) = x_0$ gives $A\cos\phi = x_0$, while $x'(0) = v_0$ gives $v_0 = A\alpha\cos\phi - \beta A\sin\phi$. By using the first of these equations in the second we find $A\sin\phi = (\alpha x_0 - v_0)/\beta$. The identity $A^2 = A^2(\cos^2\phi + \sin^2\phi) = \left[x_0^2 + (\alpha x_0 - v_0)^2/\beta^2\right]$ gives A. The identity $\tan\phi = \frac{A\sin\phi}{A\cos\phi} = \frac{(\alpha x_0 - v_0)/\beta}{x_0}$ gives ϕ.

(a) The curves $Ae^{\alpha t}\cos(\beta t + \phi)$ represent damped oscillations with "amplitude" $Ae^{\alpha t}$, so the envelopes are the curves $\pm Ae^{\alpha t}$.

(b) The curves $\pm Ae^{\alpha t}$ and $Ae^{\alpha t}\cos(\beta t + \phi)$ touch when $\pm Ae^{\alpha t} = Ae^{\alpha t}\cos(\beta t + \phi)$, that is, when $\cos(\beta t + \phi) = \pm 1$. The tangents to $\pm Ae^{\alpha t}$ have slopes
$(\pm Ae^{\alpha t})' = \pm A\alpha e^{\alpha t}$, while the tangents to $Ae^{\alpha t}\cos(\beta t + \phi)$ have slopes
$[Ae^{\alpha t}\cos(\beta t + \phi)]' = \alpha Ae^{\alpha t}\cos(\beta t + \phi) - \beta Ae^{\alpha t}\sin(\beta t + \phi)$.
These slopes are the same when $\sin(\beta t + \phi) = 0$, that is, when $\cos(\beta t + \phi) = \pm 1$, which is when they touch.

(c) The local extrema of the curve $x(t) = Ae^{\alpha t}\cos(\beta t + \phi)$ occur when
$\alpha Ae^{\alpha t}\cos(\beta t + \phi) - \beta Ae^{\alpha t}\sin(\beta t + \phi) = 0$,
that is, when $\tan(\beta t + \phi) = \alpha/\beta$, so
$\beta t + \phi = \arctan(\alpha/\beta) + n\pi$, $n = 0, \pm 1, \pm 2, \cdots$.
When this occurs, the height of the curve is given by
$x = Ae^{\alpha t}\cos(\arctan(\alpha/\beta) + n\pi) = \pm Ae^{\alpha t}\cos(\arctan(\alpha/\beta))$,
where $t = [\arctan(\alpha/\beta) + n\pi - \phi]/\beta$. The curves $\pm Ae^{\alpha t}\cos(\arctan(\alpha/\beta))$ pass through these points.

(d) The curve that oscillates is $Ae^{\alpha t}\cos(\beta t + \phi)$. The dotted curves are tangent to it when they touch, so the dotted curves are the envelopes $\pm Ae^{\alpha t}$. The other two curves pass through the extrema of $Ae^{\alpha t}\cos(\beta t + \phi)$ and so they are the curves $\pm Ae^{\alpha t}\cos(\arctan(\alpha/\beta))$.

17. **(a)** The characteristic equation is $r^2 + 2\alpha r + \lambda^2 = 0$ with solutions $r = -\alpha \pm i\omega$, where $\omega = \sqrt{\lambda^2 - \alpha^2}$, for the underdamped case. Thus, $x(t) = e^{-\alpha t}(C_1\cos\omega t + C_2\sin\omega t)$. We also have $x'(t) = e^{-\alpha t}(-\alpha C_1\cos\omega t - \alpha C_2\sin\omega t - \omega C_1\sin\omega t + \omega C_2\cos\omega t)$, so that $x'(0) = 0$ gives $C_2 = \alpha C_1/\omega$. Also $x(0) = x_0$ gives $C_1 = x_0$, so $x(t) = x_0 e^{-\alpha t}[\cos\omega t + (\alpha/\omega)\sin\omega t]$.

(b) $x'(t) = -x_0 e^{-\alpha t}\left[(\alpha^2 + \omega^2)/\omega\right]\sin\omega t$, so $x(t)$ has local maxima and minima when $\omega t = n\pi$, in which case $x(t) = \pm x_0 e^{-\alpha t}$.

(c) The amplitude should follow $x_0 e^{-\alpha t}$.

(d) A plot of $\ln x$ versus t for the data set in Table 6.3 shows a straight line with slope -0.00475 and intercept 2.336, so $\ln x = -0.00475t + 2.336$, and so $x(t) = x_0 e^{-\alpha t}$ with $x_0 = e^{2.336} \approx 10.34$ and $\alpha = 0.0047$.

(e) The mass of pendulum is displaced from its vertical position, and released from rest. Measure the maximum amplitude as a function of time.

19. **(a)** $X = (mx + My)/(m + M)$ and $Y = y - x$ give

$X' = (mx' + My')/(m + M)$,

$Y' = y' - x'$,

$X'' = (mx'' + My'')/(m + M)$,

$Y'' = y'' - x''$.

From $mx'' = k(y - x)$ and $My'' = -k(y - x)$ we have $X'' = 0$ and

$Y'' = -k(y - x)/M - k(y - x)/m = -\omega^2 (y - x) = -\omega^2 Y$,

where $\omega^2 = \frac{k}{M} + \frac{k}{m} = \frac{k(m+M)}{mM}$.

$X'' = 0$ gives $X(t) = At + B$.

$Y'' + \omega^2 Y = 0$ gives $Y(t) = C_1 \cos \omega t + C_2 \sin \omega t$.

X is the center of mass of x and y.

(b) A frictionless spring with mass $mM/(m + M)$ is governed by $mM/(m + M)z'' + kz = 0$, that is $z'' + \omega^2 z = 0$, which is the same equation that governs Y.

(c) Solving $X = (mx + My)/(m + M)$ and $Y = y - x$, for x and y gives $x = X - MY/(m + M)$ and $y = X + mY/(m + M)$. Thus,

$x(t) = At + B - M/(m + M)(C_1 \cos \omega t + C_2 \sin \omega t)$ and

$y(t) = At + B + m/(m + M)(C_1 \cos \omega t + C_2 \sin \omega t)$.

21. The characteristic equation is $(1/4)r^2 + (1/R)r + 5 = 0$ with roots $r_{1,2} = -2/R \pm i\beta$, where $\beta = 2\sqrt{5 - 1/R^2}$. The solution is

$V(t) = e^{-2t/R}(C_1 \cos \beta t + C_2 \sin \beta t)$.

$V(0) = 1$ gives $C_1 = 1$. Also,

$V'(t) = (-2/R)e^{-2t/R}(C_1 \cos \beta t + C_2 \sin \beta t) + e^{-2t/R}(-\beta C_1 \sin \beta t + \beta C_2 \cos \beta t)$

so $V'(0) = 0$ gives $-2/R + \beta C_2 = 0$, so $C_2 = 2/(\beta R)$ and

$V(t) = e^{-2t/R}(\cos \beta t + 2/(\beta R) \sin \beta t)$.

23. $(1/9)r^2 + (1/R)r + (1/4) = 0$ gives $r = \left(-1/R \pm \sqrt{1/R^2 - 1/9}\right)/(2/9)$.

(a) Overdamped if $1/R^2 - 1/9 > 0$, that is, $9 > R^2$, or $3 > R > 0$.

(b) Underdamped if $1/R^2 - 1/9 < 0$, that is, $9 < R^2$, or $R > 3$.

(c) Critically damped if $1/R^2 - 1/9 = 0$, that is, $9 = R^2$, or $R = 3$.

25. The characteristic equation is $(1/4)r^2 + r + 5 = 0$ with roots $r_{1,2} = -2 \pm 4i$. The solution is

$V(t) = e^{-2t}(C_1 \cos 4t + C_2 \sin 4t)$.

$V(0) = 0$ gives $C_1 = 0$, so

$V'(t) = -2e^{-2t}C_2 \sin 4t + 4e^{-2t}C_2 \cos 4t$ so $V'(0) = 1$ gives $C_2 = 1/4$ and

$V(t) = \frac{1}{4}e^{-2t} \sin 4t$.

The maximum value of $V(t)$ for $t \geq 0$ occurs when $V'(t) = 0$ the first time, namely $t = (\arctan 2)/4$, in which case $V = \exp\left(-\frac{1}{2}\arctan 2\right)/\left(2\sqrt{5}\right) \approx 0.12855$. The settling time is the time T after which $|V(T)| \leq 0.01 \times 0.12855$. If we plot $\frac{1}{4}e^{-2t} \sin 4t$ and the horizontal lines $\pm 0.01 \times 0.12855$, we see that $-0.01 \times 0.12855 \leq \frac{1}{4}e^{-2t} \sin 4t \leq 0.01 \times 0.12855$ from somewhere between $t = 2.2$ and $t = 2.3$. Thus, we solve $\frac{1}{4}e^{-2T} \sin 4T = \pm 0.01 \times 0.12855$ for T between 2.1 and 2.2, and we find $T \approx 2.2388$.

6.5 Interpreting the Phase Plane

1. **(a)** $x'' + 4x = 0$. With $x = y'$ we have $y' = -4x$, so $\frac{dy}{dx} = \frac{y'}{x'} = \frac{-4x}{y}$.

(b) At $t = 0$ we have $x(0) = 2$ and $x'(0) = 0$, which are the initial conditions for $x'' + 4x = 0$. In the phase plane the initial conditions for $\frac{dy}{dx} = \frac{-4x}{y}$ are $x = 2$, $y = 0$.

(c) Integrating $\frac{dy}{dx} = \frac{-4x}{y}$ gives the ellipses $\frac{1}{2}y^2 = -2x^2 + C$. From $x = 2$, $y = 0$ we have $C = 8$, so the orbits are $4x^2 + y^2 = 16$.

(d) $x'' + 4x = 0$ has characteristic equation $r^2 + 4 = 0$, with solutions $r = \pm 2i$, so that $x(t) = C_1 \cos 2t + C_2 \sin 2t$. We thus have $x'(t) = -2C_1 \sin 2t + 2C_2 \cos 2t$. Finally, $x'(0) = 0$ gives $C_2 = 0$ and $x(0) = 2$ gives $C_1 = 2$, so that $x(t) = 2\cos 2t$.

3. **(a)** From $16x'' + x = 0$ we have $16x''x' + xx' = 0$, so that $8(x')^2 + \frac{1}{2}x^2 = E$. But $x(0) = -2$ and $x'(0) = 3$ so that $8 \cdot 3^2 + \frac{1}{2} \cdot (-2)^2 = E$, that is, $E = 74$.

(b) The potential energy is $\frac{1}{2}x^2$ and the kinetic energy is $8(x')^2$. Because the sum of these two is a constant, when one is at a maximum the other must be at a minimum. In the phase plane the orbit is given by $8y^2 + \frac{1}{2}x^2 = 74$, which is an ellipse. The potential energy is a minimum when $x = 0$ in which case $y = \pm\sqrt{37/4}$. These occur at the highest and lowest points on the ellipse.

5. The differential equation in the phase plane is $dy/dx = (-by - kx)/(my)$. Horizontal tangents occur along the line $y = -kx/b$ if $b \neq 0$ and $x = 0$ if $b = 0$.

7. $x'' = -g + ce^{-ax}(x')^2$ is a second order, nonlinear, autonomous equation. $x' = y$, $y' = -g + ce^{-ax}y^2$ is a nonlinear, autonomous system of equations.

$$\frac{dy}{dx} = \frac{y'}{x'} = \frac{-g + ce^{-ax}y^2}{y} = -\frac{g}{y} + ce^{-ax}y.$$

To solve this Bernoulli equation, we write it as

$$y\frac{dy}{dx} - ce^{-ax}y^2 = -g$$

and change variable to $u = y^2$ so that $\frac{du}{dx} = 2y\frac{dy}{dx}$ and the Bernoulli equation becomes the linear equation

$$\frac{du}{dx} - 2ce^{-ax}u = -2g$$

with integrating factor $\mu = e^{\alpha}$, where $\alpha = \int (-2ce^{-ax})\,dx = 2\frac{c}{a}e^{-ax}$, so that

$$\frac{d}{dx}(e^{\alpha}u) = -2ge^{\alpha}.$$

Thus,

$$e^{\alpha}u = -2g\int e^{\alpha}\,dx + C,$$

or

$$y^2(x) = -2ge^{-\alpha}\int e^{\alpha}\,dx + Ce^{-\alpha}.$$

At time $t = 0$ we have $x = h$ and $y = 0$, so

$$y^2(x) = -2ge^{-\alpha}\int_h^x e^{\alpha}\,du = 2ge^{-\alpha}\int_x^h e^{\alpha}\,du.$$

9. **(a)** Because $\frac{1}{C} = \frac{R^2}{4L}$, the characteristic equation, $Lr^2 + Rr + \frac{R^2}{4L} = 0$, has a repeated root $r = -\frac{R}{2L}$ and so
$x(t) = (C_1 + C_2 t)e^{rt}$. Because $x'(t) = (rC_1 + rC_2 t + C_2)e^{rt}$, the initial conditions $x(0) = x_0$, $x'(0) = 0$ give $C_2 = -rC_1$, and $C_1 = x_0$. Thus,
$x(t) = x_0(1 - rt)e^{rt}$.

(b) With $R = 4$, $L = 1$ and $C = \frac{1}{4}$, we have $R^2 = 4\frac{L}{C}$ so $r = -2$ and $x(t) = x_0(1 + 2t)e^{-2t}$. The maximum of x will occur when $x' = 0$, that is, when $t = 0$. The time T at which $x(T) = 0.01x(0)$ will satisfy $(1 + 2T)e^{-2T} = 0.01$. Solving this numerically, we find $T \approx 3.319$.

11. $y' = \frac{-ay - 9x/4}{y}$ is a first order differential equation with homogeneous coefficients. Changing variables $y = xz$ gives $xz' + z = (-az - 9/4)/z$, or

$$xz' = -\frac{z^2 + az + 9/4}{z}.$$

(i) $a = 1$.

$$xz' = -\frac{z^2 + z + 9/4}{z}.$$

This has no equilibrium solutions.

The nonequilbrium solutions are obtained from

$$\int \frac{z}{z^2 + z + 9/4}\,dz = -\int \frac{1}{x}\,dx.$$

But $z^2 + z + 9/4 = \left(z + \frac{1}{2}\right)^2 + 2$, so

$$\int \frac{z}{z^2 + z + 9/4}\,dz = \int \frac{z}{(z + 1/2)^2 + 2}\,dz.$$

Changing variables by $z + \frac{1}{2} = \sqrt{2}\tan u$ gives

$$\int \frac{z}{(z + 1/2)^2 + 2}\,dz = \int \left[\tan u - 1/\left(2\sqrt{2}\right)\right]du = -\ln|\cos u| - u/\left(2\sqrt{2}\right) + C.$$

Thus,

$$-\ln|x| = \frac{1}{2}\ln\left[\left(\frac{y}{x} + \frac{1}{2}\right)^2 + 2\right] - \frac{1}{2\sqrt{2}}\arctan\left(\frac{\frac{y}{x} + \frac{1}{2}}{\sqrt{2}}\right) + C.$$

(ii) $a = 3$

$$xz' = -\frac{(z + 3/2)^2}{z}.$$

This has the equilibrium solution $z = -\frac{3}{2}$, that is $y(x) = -\frac{3}{2}x$.

The nonequilbrium solutions are obtained from

$$\int \frac{z}{(z + 3/2)^2}\,dz = -\int \frac{1}{x}\,dx.$$

Changing variables by $z + 3/2 = u$ gives

$$\int \frac{u - 3/2}{u^2}\,du = \ln|u| + \frac{3}{2}u^{-1} + C.$$

Thus,

$$-\ln|x| = \ln\left|\frac{y}{x}+\frac{3}{2}\right| + \frac{3}{2}\left(\frac{y}{x}+\frac{3}{2}\right)^{-1} + C.$$

(iii) $a = 5$.

$$xz' = -\frac{z^2+5z+9/4}{z}.$$

Because $z^2+5z+9/4 = (z+1/2)(z+9/2)$, this has two equilibrium solutions, $z = -1/2$ and $z = -9/2$, that is $y(x) = -x/2$ and $y(x) = -9x/2$.

The nonequilbrium solutions are obtained from

$$\int \frac{z}{(z+1/2)(z+9/2)}\,dz = -\int \frac{1}{x}\,dx.$$

Using partial fractions we find

$$\frac{z}{(z+1/2)(z+9/2)} = -\frac{1}{8}\frac{1}{z+1/2} + \frac{9}{8}\frac{1}{z+9/2},$$

so

$$\int \frac{z}{(z+1/2)(z+9/2)}\,dz = -\frac{1}{8}\ln|z+1/2| + \frac{9}{8}\ln|z+9/2| + C.$$

Thus,

$$-\ln|x| = -\frac{1}{8}\ln\left|\frac{y}{x}+1/2\right| + \frac{9}{8}\ln\left|\frac{y}{x}+9/2\right| + C.$$

6.6 How Explicit Solutions Are Related to Orbits

1. **(a)** The equilibrium point is $(0,0)$.

(c) The equilibrium point is $(3,9)$.

(e) This is not an autonomous system of differential equations.

3. **(a)** $x' = 0$ requires $2y = 6$, that is $y = 3$, and $y' = 0$ requires $-2x + 7y = 9$, so that $2x = 7y - 9 = 21 - 9 = 12$. Equilibrium point is $(6,3)$

(c) $x' = 0$ requires $x - 3y = 2$, and $y' = 0$ requires $-2x + 6y = -4$ or $-x + 3y = -2$. Solving these gives $x - 3y = 2$, so there are an infinite number of equilibrium points at $(2+3y, y)$ for every y.

5. **(i)** **(a)** $y = 2x - x'$, so $y' = -x + 2y$ becomes $2x' - x'' = -x + 2(2x - x')$, or
$x'' - 4x' + 3x = 0$
with general solution
$x(t) = C_1e^{3t} + C_2e^t$.

(b) From $y = 2x - x'$ we have
$y(t) = 2\left(C_1e^{3t} + C_2e^t\right) - \left(3C_1e^{3t} + C_2e^t\right)$
$= -C_1e^{3t} + C_2e^t$.
$x(0) = C_1 + C_2 = 2$
$y(0) = -C_1 + C_2 = 0$
so $C_1 = C_2 = 1$, and
$x(t) = e^{3t} + e^t$
$y(t) = -e^{3t} + e^t$.

(c) Adding and subtracting these equations gives
$x + y = 2e^t$ and $x - y = 2e^{3t}$. Thus,
$x - y = 2\left(e^t\right)^3 = (x-y)^3/4$.

(iii) (a) $y = (3x - x')/2$, so $y' = 2x - 2y$ becomes $(3x' - x'')/2 = 2x - (3x - x')$, or
$x'' - x' - 2x = 0$
with general solution
$x(t) = C_1 e^{2t} + C_2 e^{-t}$.

(b) From $y = (3x - x')/2$ we have
$y(t) = 3\left(C_1 e^{2t} + C_2 e^{-t}\right)/2 - \left(2C_1 e^{2t} - C_2 e^{-t}\right)/2$
$= C_1 e^{2t}/2 + 2C_2 e^{-t}$.
$x(0) = C_1 + C_2 = 2$
$y(0) = C_1/2 + 2C_2 = 1$
so $C_1 = 2$, $C_2 = 0$, and
$x(t) = 2e^{2t}$
$y(t) = e^{2t}$.

(c) $x = 2e^{2t} = 2y$.

6.7 The Motion of a Nonlinear Pendulum

1. (a) These occur at the equilibrium points $(2n\pi, 0)$, $n = 0, \pm 1, \pm 2, \cdots$. There are an infinite number of these, although only three occur on Figure 6.41 of the text.

   (b) These occur at the equilibrium points $((2n+1)\pi, 0)$, $n = 0, \pm 1, \pm 2, \cdots$. There are an infinite number of these, although only four occur on Figure 6.41 of the text.

3. (a) With $y = x'$, the orbit in the phase plane satisfies $\frac{dy}{dx} = -\frac{\sin x}{y}$, with solution $\frac{1}{2}y^2 = \cos x + C$. This can be rewritten as $y^2 + 4\sin^2 \frac{x}{2} = c$. The initial conditions are $x = 0$, $y = \beta$, so that $c = \beta^2$, and this orbit is $y^2 + 4\sin^2 \frac{x}{2} = \beta^2$, so $y = \pm\sqrt{\beta^2 - 4\sin^2 \frac{x}{2}}$. For x to be bounded we want periodic solutions, so if β satisfies $0 \le \beta^2 \le 4$, this will happen.

   (b) Unbounded x will occur when $\beta^2 > 4$.

   (c) No, because with this initial condition, $y^2 + 4\sin^2 \frac{x}{2} = 4\sin^2 \frac{\alpha}{2}$, and all motion is periodic.

5. (a) $dy/dx = \lambda^2(x^3/6 - x)/y$. This has three equilibrium points, namely, $(0,0)$, $\left(\sqrt{6}, 0\right)$, and $\left(-\sqrt{6}, 0\right)$, compared to only one in the linearized case.

   (b) $dy/dx = \lambda^2(x^3/6 - x)/y$, so integration gives $y^2/2 = \lambda^2(x^4/24 - x^2/2) + C$, in contrast to $y^2/2 = -\lambda^2 x^2/2 + C$.

   (c) $dy/dx = \lambda^2(x^3/6 - x)/y$. This has three equilibrium points, namely, $(0,0)$, $\left(\sqrt{6}, 0\right)$, and $\left(-\sqrt{6}, 0\right)$, compared to an infinite number in the nonlinear case.

   (d) If in $y^2 - 2\lambda^2 \cos x = c$, $\cos x$ is expanded in a Taylor series about zero, you have
   $y^2 - 2\lambda^2(1 - x^2/2 + x^4/24 - \cdots) = c$.
   If the term $-2\lambda^2$ is combined with the constant on the right hand side of the equation, the remaining two terms are identical with those given by the equation of the trajectories given in part (b).

7. (a) The period is unchanged.

   (b) The period gets larger as the amplitude is increased.

   (c) The period gets smaller as the amplitude is increased.

7. SECOND ORDER LINEAR DIFFERENTIAL EQUATIONS WITH FORCING FUNCTIONS

7.1 The General Solution

1. (a) $x_p = -2 - t^2$, $x_p' = -2t$, $x_p'' = -2$. Thus,
   $x_p'' - x_p = -2 - \left(-2 - t^2\right) = t^2$.
   The general solution of the associated homogeneous equation $x'' - x = 0$ is
   $x_h = C_1 e^t + C_2 e^{-t}$ and so the general solution of $x'' - x = t^2$ is
   $x(t) = x_h(t) + x_p(t) = C_1 e^t + C_2 e^{-t} - 2 - t^2$.

   (c) $x_p = te^{-t}$, $x_p' = -te^{-t} + e^{-t}$, $x_p'' = te^{-t} - 2e^{-t}$. Thus,
   $x_p'' + 3x_p' + 2x_p = te^{-t} - 2e^{-t} + 3\left(-te^{-t} + e^{-t}\right) + 2te^{-t} = e^{-t}$.
   The general solution of the associated homogeneous equation $x'' + 3x' + 2x = 0$ is
   $x_h = C_1 e^{-2t} + C_2 e^{-t}$ and so the general solution of $x'' + 3x' + 2x = e^{-t}$ is
   $x(t) = C_1 e^{-2t} + C_2 e^{-t} + te^{-t}$.

   (e) $x_p = 2t^3 e^{2t}$, $x_p' = 4t^3 e^{2t} + 6t^2 e^{2t}$, $x_p'' = 8t^3 e^{2t} + 24t^2 e^{2t} + 12te^{2t}$. Thus,
   $x_p'' - 4x_p' + 4x_p = 8t^3 e^{2t} + 24t^2 e^{2t} + 12te^{2t} - 4\left(4t^3 e^{2t} + 6t^2 e^{2t}\right) + 4\left(2t^3 e^{2t}\right) = 12te^{2t}$.
   The general solution of the associated homogeneous equation $x'' - 4x' + 4x = 0$ is
   $x_h = C_1 e^{2t} + C_2 te^{2t}$ and so the general solution of $x'' - 4x' + 4x = 12te^{2t}$ is
   $x(t) = C_1 e^{2t} + C_2 te^{2t} + 2t^3 e^{2t}$.

3. $x_p(t) = t \Longrightarrow x_p' = 1 \Longrightarrow x_p'' = 0 \Longrightarrow x_p'' + x_p' = 0 + 1 = 1$.
   $x_p(t) = t + 2 \Longrightarrow x_p' = 1 \Longrightarrow x_p'' = 0 \Longrightarrow x_p'' + x_p' = 0 + 1 = 1$.

   (a) The general solution of the associated homogeneous equation $x'' + x' = 0$ is
   $x_h = C_1 + C_2 e^{-t}$ and so the general solution of $x'' + x' = 1$ is
   $x(t) = x_h(t) + x_p(t) = C_1 + C_2 e^{-t} + t$.

   (b) The general solution of the associated homogeneous equation $x'' + x' = 0$ is
   $x_h = C_1 + C_2 e^{-t}$ and so the general solution of $x'' + x' = 1$ is
   $x(t) = x_h(t) + x_p(t) = C_1 + C_2 e^{-t} + t + 2$.

   (c) The constant 2 in the solution in part (b) may be absorbed into C_1 giving the solution in part (a).

5. (b) As time progresses the solution has the general shape of the forcing function. $x(t)/\left(20e^{3t}\right) = \left(e^{-2t} + e^{-t} + e^{3t}\right)/\left(20e^{3t}\right) \to 1/20$ as $t \to \infty$. This says that, as time progresses, the output function is 1/20 of the forcing function, which is consistent with the previous observation.

   (c) As time progresses the solution has the general shape of the forcing function. $x(t)/e^{-t} = \left(e^{-2t} + e^{-t} + te^{-t}\right)/e^{-t} \to \infty$ (like $t+1$) as $t \to \infty$. This says that, as time progresses, the output function is $t+1$ times the forcing function. However, both deay to zero as $t \to \infty$, which is consistent with the previous observation.

(e) As time progresses the solution has the general shape of the forcing function. $x(t)/\left(12te^{2t}\right) = \left(e^{2t} + te^{2t} + 2t^3e^{2t}\right)/\left(12te^{2t}\right) \to \infty$ [like $\left(1+2t^2\right)/12$] as $t \to \infty$. This says that, as time progresses, the output function is $\left(1+2t^2\right)/12$ times the forcing function. However, both deay to zero as $t \to \infty$, which is consistent with the previous observation.

7.2 Finding Solutions by the Method of Undetermined Coefficients

1. (a) The associated homogeneous equation is $x'' - x = 0$ with solution
$x_h(t) = C_1e^t + C_2e^{-t}$.
A trial solution corresponding to t^2 is
$x_p(t) = At^2 + Bt + C$, and this is not contained in x_h.
$x_p' = 2At + B$, $x_p'' = 2A$, so
$x_p'' - x_p = 2A - At^2 - Bt - C = t^2$ requires
$-A = 1$, $-B = 0$, and $2A - C = 0$, that is
$A = -1$, $B = 0$, and $C = -2$. Thus,
$x_p = -t^2 - 2$ and
$x(t) = C_1e^t + C_2e^{-t} - t^2 - 2$. This agrees with Exercise 1 in Section 7.1.

(c) The associated homogeneous equation is $x'' + 3x' + 2x = 0$ with solution
$x_h(t) = C_1e^{-2t} + C_2e^{-t}$.
A trial solution corresponding to e^{-t} is Ae^{-t}, but this is contained in x_h, so we try $x_p(t) = Ate^{-t}$, which is not contained in x_h.
$x_p' = -Ate^{-t} + Ae^{-t}$, $x_p'' = Ate^{-t} - 2Ae^{-t}$, so
$x_p'' + 3x_p' + 2x_p = Ate^{-t} - 2Ae^{-t} + 3\left(-Ate^{-t} + Ae^{-t}\right) + 2Ate^{-t} = e^{-t}$ requires $A = 1$. Thus,
$x_p = te^{-t}$ and
$x(t) = C_1e^{-2t} + C_2e^{-t} + te^{-t}$. This agrees with Exercise 1 in Section 7.1.

(e) The associated homogeneous equation is $x'' - 4x' + 4x = 0$ with solution
$x_h(t) = C_1e^{2t} + C_2te^{2t}$.
A trial solution corresponding to $12te^{2t}$ is $(At+B)e^{2t}$, but this is contained in x_h, so we try $\left(At^2 + Bt\right)e^{2t}$, but this is also contained in x_h so we try
$x_p(t) = \left(At^3 + Bt^2\right)e^{2t}$ which is not contained in x_h.
$x_p' = \left[2At^3 + (2B+3A)t^2 + 2Bt\right]e^{2t}$,
$x_p'' = \left[4At^3 + 2(2B+6A)t^2 + 2(4B+3A)t + 2B\right]e^{2t}$, so
$x_p'' - 4x_p' + 4x_p$
$= \left[4At^3 + 2(2B+6A)t^2 + 2(4B+3A)t + 2B\right]e^{2t} +$
$\quad -4\left[2At^3 + (2B+3A)t^2 + 2Bt\right]e^{2t} + 4\left(At^3 + Bt^2\right)e^{2t}$
$= (6At + 2B)e^{2t}$
$= 12te^{2t}$ requires
$6A = 12$ and $B = 0$. Thus,
$x_p = 2t^3e^{2t}$ and $x(t) = C_1e^{2t} + C_2te^{2t} + 2t^3e^{2t}$. This agrees with Exercise 1 in Section 7.1.

3. In all cases the associated homogeneous equation is $x'' + x' - 6x = 0$ with solution $x_h(t) = C_1e^{2t} + C_2e^{-3t}$.

(a) The forcing term $28e^{4t}$ is not a solution of the associated homogeneous equation so our trial solution is $x_p(t) = Ae^{4t}$.
Thus, $x_p' = 4Ae^{4t}$ and $x_p'' = 16Ae^{4t}$ so
$x_p'' + x_p' - 6x_p = (16 + 4 - 6)\, Ae^{4t} = 14Ae^{4t}$.
For this to be $28e^{4t}$ we need $A = 2$ so $x_p(t) = 2e^{4t}$ and the general solution is
$x(t) = x_h(t) + x_p(t) = C_1e^{2t} + C_2e^{-3t} + 2e^{4t}$.

(c) The forcing term $8e^{2t}$ is a solution of the associated homogeneous equation so our trial solution is $x_p(t) = Ate^{2t}$ which is not a solution of the associated homogeneous equation.
Thus, $x_p' = Ae^{2t}(2t + 1)$ and $x_p'' = Ae^{2t}(4t + 4)$ so
$x_p'' + x_p' - 6x_p = (4t + 4 + 2t + 1 - 6t)\, Ae^{2t} = 5Ae^{2t}$.
For this to be $8e^{2t}$ we need $A = 8/5$ so $x_p(t) = 8/5te^{2t}$ and the general solution is
$x(t) = x_h(t) + x_p(t) = C_1e^{2t} + C_2e^{-3t} + 8/5te^{2t}$.

(e) $\sin t^2$ is not in Table 7.1.

5. We list the solution of the associated homogeneous equation, the trial particular solution, and then the solution.

(a) $x_h(t) = C_1e^{2t} + C_2e^{-3t}$,
$x_p(t) = At + B$,
$x(t) = -\frac{1}{36} - \frac{1}{6}t + C_1e^{2t} + C_2e^{-3t}$.

(c) $x_h(t) = C_1e^{2t} + C_2e^{-3t}$,
$x_p(t) = A\sin t + B\cos t$,
$x(t) = -\frac{7}{50}\sin t - \frac{1}{50}\cos t + C_1e^{2t} + C_2e^{-3t}$.

(e) $x_h(t) = C_1 + C_2e^{-t}$,
The trial solution Ae^{-t} is contained in x_h, so we multiply this guess by t, obtaining $x_p(t) = Ate^{-t}$, which is not in x_h.
$x(t) = -6te^{-t} + C_1 + C_2e^{-t}$.

(g) $x_h(t) = C_1\cos 2t + C_2\sin 2t$,
$x_p(t) = A\cos 3t + B\sin 3t$,
$x(t) = -2\sin 3t + C_1\cos 2t + C_2\sin 2t$.

(i) $x_h(t) = C_1\cos 2t + C_2\sin 2t$,
The trial solution $A\cos 2t + B\sin 2t$ is contained in x_h, so we multiply this guess by t, obtaining
$x_p(t) = t\,(A\cos 2t + B\sin 2t)$, which is not in x_h.
$x(t) = -\frac{1}{4}t\cos 2t + C_1\cos 2t + C_2\sin 2t$.

(k) $x_h(t) = C_1e^{\frac{3}{2}t} + C_2te^{\frac{3}{2}t}$,
The trial solution $(At + B)\, e^{\frac{3}{2}t}$ is contained in x_h, so we multiply this guess by t, obtaining $t\,(At + B)\, e^{\frac{3}{2}t}$ which is also in x_h, so we multiply this guess by t, obtaining
$x_p(t) = t^2\,(At + B)\, e^{\frac{3}{2}t}$, which is not in x_h.
$x(t) = t^3e^{\frac{3}{2}t} + C_1e^{\frac{3}{2}t} + C_2te^{\frac{3}{2}t}$.

7. (a) The solution of $x'' + x' - 12x = 8e^{3t}$ is
$x(t) = C_1e^{3t} + C_2e^{-4t} + \frac{8}{7}te^{3t}$.
The condition $x(0) = 0$ gives $0 = C_1 + C_2$, so $C_2 = -C_1$.
From $x'(t) = 3C_1e^{3t} - 4C_2e^{-4t} + \frac{8}{7}e^{3t} + \frac{24}{7}te^{3t}$ and $x'(0) = 1$, we also find $1 = 3C_1 - 4C_2 + \frac{8}{7}$, or $7C_1 = -\frac{1}{7}$. Thus, $C_1 = -\frac{1}{49}$ and $C_2 = \frac{1}{49}$, giving the solution
$x(t) = -\frac{1}{49}e^{3t} + \frac{8}{7}te^{3t} + \frac{1}{49}e^{-4t}$.

(c) The solution of $x'' - 5x' + 6x = 12te^{-t} - 7e^{-t}$ is
$x(t) = C_1 e^{3t} + C_2 e^{2t} + te^{-t}$.
The condition $x(0) = 0$ gives $0 = C_1 + C_2$, so $C_2 = -C_1$.
From $x'(t) = 3C_1 e^{3t} + 2C_2 e^{2t} - te^{-t} + e^{-t}$ and $x'(0) = 0$, we also find
$0 = 3C_1 + 2C_2 + 1$, or $0 = 3C_1 - 2C_1 + 1$. Thus, $C_1 = -1$ and $C_2 = 1$, giving the solution
$x(t) = -e^{3t} + e^{2t} + te^{-t}$.

9. $x_p = \alpha \cos at + \beta \sin at$, where $\alpha = A/\left(\omega^2 - a^2\right)$ and $\beta = B/\left(\omega^2 - a^2\right)$ gives
$x_p' = -\alpha a \sin at + a\beta \cos at$ and $x_p'' = -\alpha a^2 \cos at - \beta a^2 \sin at$, so that
$x_p'' + \omega^2 x_p = \left(\omega^2 \alpha - \alpha a^2\right) \cos at + \left(\omega^2 \beta - \beta a^2\right) \sin at = A \cos at + B \sin at$.

11. **(a)** The associated homogeneous equation is $x'' - x' = 0$ with solution
$x_h(t) = C_1 + C_2 e^t$.
A trial solution corresponding to t^2 is $At^2 + Bt + C$, but this is contained in x_h, so we try
$x_p(t) = At^3 + Bt^2 + Ct$ which is not contained in x_h.
$x_p' = 3At^2 + 2Bt + C$, $x_p'' = 6At + 2B$, so
$x_p'' - x_p' = 6At + 2B - 3At^2 - 2Bt - C = t^2$ requires
$-3A = 1$, $6A - 2B = 0$, and $2B - C = 0$, that is
$A = -1/3$, $B = -1$, and $C = -2$. Thus,
$x_p = -t^3/3 - t^2 - 2t$ and $x(t) = C_1 + C_2 e^t - t^3/3 - t^2 - 2t$.

(b) With $y = x'$ we have $y' - y = t^2$, a linear differential equation with integrating factor $\mu = e^{-t}$, so that
$e^{-t}y' - e^{-t}y = e^{-t}t^2$, or $(e^{-t}y)' = e^{-t}t^2$. Integrating gives
$e^{-t}y = \int e^{-t}t^2\,dt$. Integrating by parts twice gives
$\int e^{-t}t^2\,dt = -t^2e^{-t} + \int 2te^{-t}\,dt = -t^2e^{-t} + 2\left(-te^{-t} - e^{-t}\right)$ so
$e^{-t}y = -t^2e^{-t} + 2\left(-te^{-t} - e^{-t}\right) + C_1$, or
$x' = y = -t^2 - 2t - 2 + C_1 e^t$, so
$x(t) = -t^3/3 - t^2 - 2t + C_1 e^t + C_2$.

13. Because there is no term containing x' the trial solution $x_p = A \sin \omega t$ gives $x_p'' + a^2 x_p = -\omega^2 A \sin \omega t + a^2 A \sin \omega t$ which we can make equal to $\sin \omega t$ by choosing $A = 1/\left(-\omega^2 + a^2\right)$, as long as $\omega \neq \pm a$.

15. $x_p(t) = \sin t \ln|\csc t - \cot t|$ gives
$x_p'(t) = \cos t \ln|\csc t - \cot t| + \sin t\left(-\csc t \cot t + \csc^2 t\right)/(\csc t - \cot t)$
$= \cos t \ln|\csc t - \cot t| + 1$
$x_p''(t) = -\sin t \ln|\csc t - \cot t| + \cos t/\sin t$. Thus,
$x_p'' + x_p(t) = \cot t$.
$x_h'' + x_h = 0$ has solution $x_h = C_1 \cos t + C_2 \sin t$, so
$x(t) = C_1 \cos t + C_2 \sin t + \sin t \ln|\csc t - \cot t|$.

17. $a_2 (ax_1)'' + a_1 (ax_1)' + a_0 (ax_1)$
$= aa_2 x_1'' + aa_1 x_1' + aa_0 x_1$
$= a\left(a_2 x_1'' + a_1 x_1' + a_0 x_1\right)$
$= af(t)$.

19. $a_2(x_1+x_2)''+a_1(x_1+x_2)'+a_0(x_1+x_2)$

$= (a_2x_1''+a_1x_1'+a_0x_1)+(a_2x_2''+a_1x_2'+a_0x_2)$

$= f_1(t)+f_2(t)$, subject to

$x_1(t_0)+x_2(t_0)=h_1+h_2$ and $x_1'(t_0)+x_2'(t_0)=k_1+k_2$.

21. $x=e^t$ gives $x'=e^t$ and $x''=e^t$ so $xx''-2(x')^2+x^2=e^te^t-2(e^t)^2+(e^t)^2=0$.

$x=e^{-t}$ gives $x'=-e^{-t}$ and $x''=e^{-t}$ so $xx''-2(x')^2+x^2=e^{-t}e^{-t}-2(-e^{-t})^2+(e^{-t})^2=0$.

$x=ae^t+be^{-t}$ gives $x'=ae^t-be^{-t}$ and $x''=ae^t+be^{-t}$ so

$xx''-2(x')^2+x^2$

$=(ae^t+be^{-t})(ae^t+be^{-t})-2(ae^t-be^{-t})^2+(ae^t+be^{-t})^2$

$=8ab$.

7.3 Applications and Models

1. $16x''+8x'+10x=\sin\omega t$ has homogeneous equation $16x''+8x'+10x=0$ with characteristic equation $8r^2+4r+5=0$, roots $r=-\frac{1}{4}\pm\frac{3}{4}i$. Thus,

$x_h(t)=e^{-t/4}[C_1\cos(3t/4)+C_2\sin(3t/4)]$.

$x_p(t)=A\sin\omega t+B\cos\omega t\implies$

$x_p'(t)=A\omega\cos\omega t-B\omega\sin\omega t\implies$

$x_p''(t)=-A\omega^2\sin\omega t-B\omega^2\cos\omega t$, we find

$16x_p''+8x_p'+10x_p$

$=16(-A\omega^2\sin\omega t-B\omega^2\cos\omega t)+8(A\omega\cos\omega t-B\omega\sin\omega t)+10(A\sin\omega t+B\cos\omega t)$

$=(-16A\omega^2-8B\omega+10A)\sin\omega t+(-16B\omega^2+8A\omega+10B)\cos\omega t$.

For this to equal $\sin\omega t$ we require

$-16A\omega^2-8B\omega+10A=1$ and $-16B\omega^2+8A\omega+10B=0$ which give

$A=(5-8\omega^2)/\alpha$, $B=-4\omega/\alpha$ where $\alpha=2[64\omega^2-64\omega-9]$.

With this choice of A and B the solution is

$x(t)=e^{-t/4}[C_1\cos(3t/4)+C_2\sin(3t/4)]+A\sin\omega t+B\cos\omega t$.

$x(0)=0$ gives $C_1=-B$. From

$x'(t)=-(3/4)e^{-t/4}C_1\sin(3t/4)-(1/4)e^{-t/4}C_1\cos(3t/4)+(3/4)e^{-t/4}C_2\cos(3t/4)+$
$-(1/4)e^{-t/4}C_2\sin(3t/4)+\omega A\cos\omega t-\omega B\sin\omega t$, and

$x'(0)=0$ we also have $-(1/4)C_1+(3/4)C_2+\omega A=0$, so

$C_2=(1/3)C_1-(4/3)\omega A=-B/3-4\omega A/3$. Thus, with the previous choices of A and B the solution is

$x(t)=e^{-t/4}[-B\cos(3t/4)+(-B/3-4\omega A/3)\sin(3t/4)]+A\sin\omega t+B\cos\omega t$.

3. $\lim_{\omega\to 10}19(\cos\omega t-\cos 10t)/(100-\omega^2)=\lim_{\omega\to 10}-19t\sin\omega t/(-2\omega)=19t\sin 10t/(20)$.

5. The solution of the associated homogeneous equation $x'' + 2x' + 17x = 0$ is

$x_h(t) = C_1e^{-t}\sin 4t + C_2e^{-t}\cos 4t.$

Starting from the particular solution

$x_p(t) = A\sin\omega t + B\cos\omega t \Longrightarrow$

$x_p'(t) = A\omega\cos\omega t - B\omega\sin\omega t \Longrightarrow$

$x_p''(t) = -A\omega^2\sin\omega t - B\omega^2\cos\omega t$, we find

$x_p'' + 2x_p' + 17x_p$

$= \left(-A\omega^2\sin\omega t - B\omega^2\cos\omega t\right) + 2\left(A\omega\cos\omega t - B\omega\sin\omega t\right) + 17\left(A\sin\omega t + B\cos\omega t\right)$

$= \left(-A\omega^2 - 2B\omega + 17A\right)\sin\omega t + \left(-B\omega^2 + 2A\omega + 17B\right)\cos\omega t.$

For this to equal $2\sin\omega t$ we require

$-A\omega^2 - 2B\omega + 17A = 2$ and $-B\omega^2 + 2A\omega + 17B = 0$ which give

$A = -2\left(\omega^2 - 17\right)\alpha,\ B = -4\omega\alpha$ where $\alpha = 1/\left[\left(\omega^2 - 17\right)^2 + 4\omega^2\right].$

With this choice of A and B the solution is

$x(t) = C_1e^{-t}\sin 4t + C_2e^{-t}\cos 4t + A\sin\omega t + B\cos\omega t.$

(a) With $x(0) = 0$ we have $0 = C_2 + B$ so $C_2 = -B$. From

$x'(t) = C_1e^{-t}\left(-\sin 4t + 4\cos 4t\right) + C_2e^{-t}\left(-\cos 4t - 4\sin 4t\right) + A\omega\cos\omega t - B\omega\sin\omega t$ and $x'(0) = 0$ we find

$0 = 4C_1 - C_2 + A\omega$, so $C_1 = \frac{1}{4}\left(-B - A\omega\right).$

Thus,

$x(t) = \frac{1}{4}\left(-B - A\omega\right)e^{-t}\sin 4t - Be^{-t}\cos 4t + A\sin\omega t + B\cos\omega t.$

(b) The steady state solution is $A\sin\omega t + B\cos\omega t$ while the transient solution is $\frac{1}{4}\left(-B - A\omega\right)e^{-t}\sin 4t - Be^{-t}\cos 4t.$

7. The solution of the associated homogeneous equation $x'' + 6x' + 22x = 0$ is

$x_h(t) = C_1e^{-3t}\sin\sqrt{13}t + C_2e^{-3t}\cos\sqrt{13}t$. Starting from the particular solution

$x_p(t) = A\sin\omega t + B\cos\omega t \Longrightarrow$

$x_p'(t) = A\omega\cos\omega t - B\omega\sin\omega t \Longrightarrow$

$x_p''(t) = -A\omega^2\sin\omega t - B\omega^2\cos\omega t$, we find

$x_p'' + 6x_p' + 22x_p$

$= \left(-A\omega^2\sin\omega t - B\omega^2\cos\omega t\right) + 6\left(A\omega\cos\omega t - B\omega\sin\omega t\right) + 22\left(A\sin\omega t + B\cos\omega t\right)$

$= \left(-A\omega^2 - 6B\omega + 22A\right)\sin\omega t + \left(-B\omega^2 + 6A\omega + 22B\right)\cos\omega t.$

For this to equal $\cos\omega t$ we require

$-A\omega^2 - 6B\omega + 22A = 0$ and $-B\omega^2 + 6A\omega + 22B = 1$ which give

$A = 6\omega\alpha,\ B = \left(22 - \omega^2\right)\alpha$ where $\alpha = 1/\left[\left(\omega^2 - 22\right)^2 + 36\omega^2\right].$

With this choice of A and B the solution is

$x(t) = C_1e^{-3t}\sin\sqrt{13}t + C_2e^{-3t}\cos\sqrt{13}t + A\sin\omega t + B\cos\omega t.$

The steady state solution is

$A\sin\omega t + B\cos\omega t = \sqrt{A^2 + B^2}\left(\frac{A}{\sqrt{A^2+B^2}}\sin\omega t + \frac{B}{\sqrt{A^2+B^2}}\cos\omega t\right)$

with amplitude $\sqrt{A^2 + B^2} = \sqrt{\alpha}$. This has an extreme value where $d\sqrt{\alpha}/d\omega = 0$, that is, when $d\alpha/d\omega = 0$.

Now,

$$d\alpha/d\omega = -\left[4\omega\left(\omega^2-22\right)+72\omega\right] / \left[\left(\omega^2-22\right)^2+36\omega^2\right]^2$$

so $d\alpha/d\omega = 0$ when $\omega = 0$, or $4\left(\omega^2-22\right)+72=0$, that is, $\omega^2=4$, so $\omega=\pm 2$. $\omega = 0$ is a local minimum while $\omega = \pm 2$ are global maxima.

9. If $\alpha = 1/\left[4\omega^2+\left(17-\omega^2\right)^2\right]$ then $\sqrt{A^2+B^2}=\sqrt{\alpha}$.

This has an extreme value where $d\sqrt{\alpha}/d\omega = 0$ that is, when $d\alpha/d\omega = 0$.

Now,

$$d\alpha/d\omega = -\left[8\omega - 4\omega\left(17-\omega^2\right)\right] / \left[4\omega^2+\left(17-\omega^2\right)^2\right]^2$$

$$= -4\omega\left(\omega^2-15\right)\alpha^2$$

so $d\alpha/d\omega = 0$ when $\omega = 0$, or $\omega^2 = 15$, so $\omega = \pm\sqrt{15}$.

$\omega = 0$ is a local minimum while $\omega = \pm\sqrt{15}$ are global maxima.

The extreme values of $\omega\sqrt{\alpha}$ occur when $d\left(\omega\sqrt{\alpha}\right)/d\omega = 0$, that is, when $2\alpha+\omega d\alpha/d\omega = 0$. Thus, $2\alpha - 4\omega^2\left(\omega^2-15\right)\alpha^2 = 0$, which can be rewritten as $1/\alpha - 2\omega^2\left(\omega^2-15\right)=0$ or $4\omega^2+\left(17-\omega^2\right)^2 - 2\omega^2\left(\omega^2-15\right)=0$.

Expanding this gives $\omega^4 = 17^2$, so $\omega = \pm\sqrt{17}$.

11. **(a)** (i) Integrating $x'' = -1$ gives $x' = -t + C_1$, and integrating again gives $x(t) = -\frac{1}{2}t^2 + C_1 t + C_2$.

(ii) The homogeneous equation has characteristic equation $r^2 = 0$, with repeated root $r = 0$, so $x_h = C_1 t + C_2$.

The trial solution corresponding to the forcing function -1 is A, which is contained in x_h. The new trial solution At is also contained in x_h, but $x_p = At^2$ is not.

$x_p' = 2At$, $x_p'' = 2A$ so $x_p'' = -1$ gives $A = -\frac{1}{2}$, $x_p = -\frac{1}{2}t^2$, and

$x(t) = -\frac{1}{2}t^2 + C_1 t + C_2$.

The initial conditions are $x(0) = 0$, $x'(0) = 1$, so $C_2 = 0$ and $C_1 = 1$, and

$x(t) = -\frac{1}{2}t^2 + t$.

Its maximum height will be attained when $x' = 0$, that is $t = 1$, so it takes 1 unit of time to reach maximum height. It returns to the ground when $x = 0$ which is at $t = 2$. Thus, it takes 1 unit to fall from its maximum height to the ground, so it takes as long to go up as it does to come down.

The upper curve in Figure 7-1 is $x(t) = -\frac{1}{2}t^2 + t$. Because the maximum is halfway between the two t intercepts, it takes the same time to go from the ground to its maximum height as it does to fall back to the ground.

(b) (i) Integrating $x'' + x' = -1$ gives $x' + x = -t + C_1$. This is a linear differential equation with integrating factor $\mu = e^t$, so

$\left(e^t x\right)' = \left(-t + C_1\right)e^t$. Integrating gives

$e^t x = -\int te^t\,dt + C_1 e^t + C_2 = -te^t + e^t + C_1 e^t + C_2$, so

$x(t) = -t + 1 + C_1 + C_2 e^{-t}$.

(ii) The homogeneous equation has characteristic equation $r^2 + r = 0$, with roots $r = 0, -1$, so

$x_h = C_1 + C_2 e^{-t}$.

The trial solution corresponding to the forcing function -1 is A, which is contained in x_h. The new trial solution $x_p = At$ is not, so $x_p' = A$, $x_p'' = 0$ and $x_p'' + x_p' = -1$ gives $A = -1$, $x_p = -t$, and

$x(t) = -t + C_1 + C_2 e^{-t}$. This is the same as the answer given in part b(i) if, in that case, we absorb the 1 into C_1.

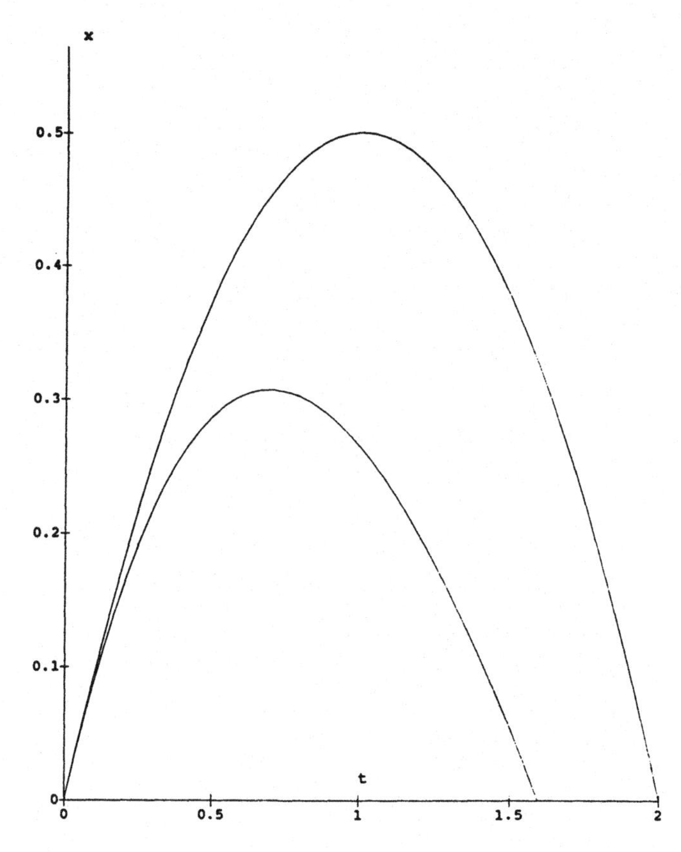

Figure 7-1 Exercise 11.

The initial conditions are $x(0) = 0$, $x'(0) = 1$, so from $x(t) = -t + C_1 + C_2 e^{-t}$ we find $C_1 + C_2 = 0$ and $-1 - C_2 = 1$, and
$x(t) = -t + 2 - 2e^{-t}$.
Its maximum height will be attained when $x' = 0$, that is when $0 = -1 + 2e^{-t}$ or $t = \ln 2$, so it takes $\ln 2 \approx 0.6931$ units of time to reach maximum height. It returns to the ground when $x = 0$ which is at time t where $-t + 2 - 2e^{-t} = 0$. Solving this numerically we find $t \approx 1.5936$. Thus, it takes about $1.5936 - 0.6931 = 0.9005$ units of time to fall from its maximum height to the ground, so it takes longer to come down than to go up.
The lower curve in Figure 7-1 is $x(t) = -t + 2 - 2e^{-t}$. Because the maximum is nearer the first of the two t intercepts, it takes less time to go from the ground to its maximum height than it does to fall back to the ground.

13. **(a)** $x'' = y' + c\cos ct = -4x + c\cos ct$.

(b) $x'' + 4x = c\cos ct$ has the solution
$x(t) = C_1 \cos 2t + C_2 \sin 2t + \frac{c}{4-c^2}\cos ct$ if $c \neq 2$, so
$y(t) = x' - \sin ct = -2C_1 \sin 2t + 2C_2 \cos 2t - \frac{4}{4-c^2}\sin ct$.
The initial conditions $x(0) = 0$, $y(0) = 0$ give
$C_1 = -\frac{c}{4-c^2}$ and $C_2 = 0$, so, if $c \neq 2$,
$x(t) = -\frac{c}{4-c^2}\cos 2t + \frac{c}{4-c^2}\cos ct = \frac{c}{4-c^2}(\cos ct - \cos 2t)$.
As $c \to 2$ the amplitude increases. At $c = 2$ the solution becomes unbounded as $t \to \infty$. As c increases for $c > 2$, the amplitude decreases.

8. SECOND ORDER LINEAR DIFFERENTIAL EQUATIONS — QUALITATIVE AND QUANTITATIVE ASPECTS

8.1 Qualitative Behavior of Solutions

1. Here $Q(t) = 1 + e^t \geq 1$ for all t, so all nontrivial solutions oscillate.

3. Here $Q(t) = \left[1 - 1/\left(4t^2\right)\right] - 1/2(1/t)' - 1/4(1/t)^2$
$= 1 - 1/\left(4t^2\right) + 1/\left(2t^2\right) - 1/\left(4t^2\right)$
$= 1 \geq 1$ for $t > 0$, so all nontrivial solutions oscillate and are bounded.
The substitution $x = uz$, where
$u = \exp\left(-\frac{1}{2}\int p\,dt\right) = \exp\left(-\frac{1}{2}\int \frac{1}{t}\,dt\right) = \exp\left(-\frac{1}{2}\ln t\right) = \frac{1}{\sqrt{t}}$
converts $x'' + \frac{1}{t}x' + \left(1 - \frac{1}{4t^2}\right)x = 0$ into $z'' + z = 0$.
The last equation has solution $z(t) = C_1 \sin t + C_2 \cos t$, so the original equation has solution $x(t) = uz = \frac{1}{\sqrt{t}}\left(C_1 \sin t + C_2 \cos t\right)$ which oscillates for $t > 0$.

5. We use part (c) of the Oscillation Theorem knowing that one solution is a polynomial of degree n which has at most n zeros. Because this solution has a finite number of zeros, so too must every other solution. Thus, no solution can oscillate.

7. **(a)** $x(t) = C_1 t^{r_1} + C_2 t^{r_2}$ where $r_{1,2} = \left(1 \pm \sqrt{1 - 4a^2}\right)/2$.
$x'(t) = r_1 C_1 t^{r_1 - 1} + r_2 C_2 t^{r_2 - 1}$,
$x''(t) = r_1\left(r_1 - 1\right) C_1 t^{r_1 - 2} + r_2\left(r_2 - 1\right) C_2 t^{r_2 - 2}$.
$x'' + a^2 x/t^2$
$= r_1\left(r_1 - 1\right) C_1 t^{r_1 - 2} + r_2\left(r_2 - 1\right) C_2 t^{r_2 - 2} + a^2\left(C_1 t^{r_1 - 2} + C_2 t^{r_2 - 2}\right)$
$= \left(r_1^2 - r_1 + a^2\right) C_1 t^{r_1 - 2} + \left(r_2^2 - r_2 + a^2\right) C_2 t^{r_2 - 2}$.
The polynomial $r^2 - r + a^2$ is zero when $r = \left(1 \pm \sqrt{1 - 4a^2}\right)/2$, so $x'' + a^2 x/t^2 = 0$.

(b) $x(t) = C_1\sqrt{t} + C_2\sqrt{t}\ln t$.
$x'(t) = (1/2) C_1 t^{-1/2} + C_2\left[(1/2)t^{-1/2}\ln t + t^{-1/2}\right]$.
$x''(t) = (-1/4) C_1 t^{-3/2} + C_2\left[(-1/4)t^{-3/2}\ln t + (1/2)t^{-3/2} - (1/2)t^{-3/2}\right]$
$= (-1/4) C_1 t^{-3/2} + C_2 t^{-3/2}\left[(-1/4)\ln t\right]$.
$x'' + 1/\left(4t^2\right)x$
$= (-1/4) C_1 t^{-3/2} + C_2 t^{-3/2}\left[(-1/4)\ln t\right] + 1/\left(4t^2\right)\left[C_1\sqrt{t} + C_2\sqrt{t}\ln t\right]$
$= 0$.

(c) $x(t) = C_1\sqrt{t}\cos(\omega \ln t) + C_2\sqrt{t}\sin(\omega \ln t)$.
$x'(t) = \frac{1}{2}C_1 t^{-1/2}\cos(\omega \ln t) - \omega C_1 t^{-1/2}\sin(\omega \ln t) +$
$+\frac{1}{2}C_2 t^{-1/2}\sin(\omega \ln t) + \omega C_2 t^{-1/2}\cos(\omega \ln t)$
$= \left(\frac{1}{2}C_1 + \omega C_2\right) t^{-1/2}\cos(\omega \ln t) + \left(\frac{1}{2}C_2 - \omega C_1\right) t^{-1/2}\sin(\omega \ln t)$

$$x''(t) = -\left(\tfrac{1}{4} + \omega^2\right) C_1 t^{-3/2} \cos(\omega \ln t) - \left(\tfrac{1}{4} + \omega^2\right) C_2 t^{-3/2} \sin(\omega \ln t)$$

$$x'' + a^2/\left(4t^2\right) x$$

$$= -\left(\tfrac{1}{4} + \omega^2 - a^2\right) C_1 t^{-3/2} \cos(\omega \ln t) - \left(\tfrac{1}{4} + \omega^2 - a^2\right) C_2 t^{-3/2} \sin(\omega \ln t)$$

$$= 0.$$

(d) The solution in part (a) can be written $x(t) = t^{r_1}\left(C_1 + C_2 t^{r_2 - r_1}\right)$.
If C_1 and C_2 have the same sign or if either is zero, $x(t)$ cannot vanish for $t > 0$.
If C_1 and C_2 have opposite signs then $x(t)$ can vanish once for $t > 0$.
The solution in part (b) can be written $x(t) = \sqrt{t}\,(C_1 + C_2 \ln t)$.
If $C_2 \neq 0$, $x(t)$ can vanish once for $t > 0$ — when $\ln t = -C_1/C_2$.
If $C_2 = 0$, $x(t)$ cannot vanish for $t > 0$.
The solution in part (c) oscillates and so has an infinite number of zeros.

9. For the first equation $Q(t) = 1 + e^{-t} \geq 1$ for all t, so all its nontrivial solutions oscillate.

11. $x(t) = t^2 - t$ gives $x' = 2t - 1$, $x'' = 2$, so

$$t^2 x'' - 2tx' + 2x = 2t^2 - 2t(2t-1) + 2\left(t^2 - t\right) = 0.$$

$x(t) = t$ gives $x' = 1$, $x'' = 0$, so

$$t^2 x'' - 2tx' + 2x = -2t + 2t = 0.$$

The Oscillation Theorem applies to differential equations where the coefficient of x'' is one. Thus we should rewrite $t^2x'' - 2tx' + 2x = 0$ in the form $x'' - \frac{2}{t}x' + \frac{2}{t^2}x = 0$ with $Q = \frac{2}{t^2} - \frac{1}{2}\left(\frac{2}{t^2}\right) - \frac{1}{4}\left(\frac{2}{t}\right)^2 = 0$. In obtaining Q we assumed $t \neq 0$, so $Q(t)$ is continuous for $t \geq t_0 > 0$. Thus, part (a) of the theorem states that any solution has at most one zero for $t > 0$. Our $x_1(t)$ has one zero, and $x_2(t)$ has no zeros for $t > 0$.Part (c) of the theorem does not apply because neither x_1 nor x_2 have two zeros for $t > 0$..

13. $x'' + 2fx' + \left(c + f' + f^2\right)x = 0$ has $Q = c + f' + f^2 - \frac{1}{2}2f' - \frac{1}{4}(2f)^2 = c$. Thus, the change of variable $x = \exp\left(-\int f\,dx\right) z$ converts the differential equation into $z'' + cz = 0$. Depending on whether $c = 0$, $c > 0$, or $c < 0$, we can solve this for z and then find x.

(a) $x'' + 2tx' + \left(1 + t^2\right)x = 0$. Put $f(t) = t$ so that $f' + f^2 = 1 + t^2$, and $c = 0$. Thus $z'' = 0$, so that $z = C_1 + C_2 t$ and
$x(t) = e^{-t^2/2}\left(C_1 + C_2 t\right).$

(c) $x'' + \frac{2}{t}x' + 4x = 0$. Put $f(t) = \frac{1}{t}$ so that $f' + f^2 = 0$, and $c = 4$. Thus, $z'' + 4z = 0$, so that $z = C_1 \cos 2t + C_2 \sin 2t$ and
$x(t) = \frac{1}{t}\left(C_1 \cos 2t + C_2 \sin 2t\right).$

15. This means that all solutions are concave up everywhere, and so they cannot oscillate.

8.2 Finding Solutions by Reduction of Order

1. **(a)** $x = e^{2t} \Longrightarrow x' = 2e^{2t} \Longrightarrow \;' = 4e^{2t} \Longrightarrow$
$x'' - 4x' + 4x = 4e^{2t} - \;(2e^{2t}) + 4\left(e^{2t}\right) = 0$, so e^{2t} is a solution.
$x = e^{2t}z \Longrightarrow x' = \;(2z + z') \Longrightarrow x'' = e^{2t}\left(4z + 4z' + z''\right)$.
Thus,
$x'' - 4x' + 4x = e^{2t}\left(4z + 4z' + z''\right) - 4e^{2t}\left(2z + z'\right) + 4e^{2t}z = 0$

reduces to

$e^{2t}z'' = 0$, or $z'' = 0$.

This gives $z = C_1 + C_2 t$, so

$x(t) = e^{2t}(C_1 + C_2 t)$ is the solution of $x'' - 4x' + 4x = 0$.

This solution has at most one zero. If we apply the Oscillation and Relation Theorems, we find $Q(t) = 4 - (1/4)\,4^2 = 0$, which requires a maximum of one zero.

(c) $x = t \Longrightarrow x' = 1 \Longrightarrow x'' = 0 \Longrightarrow$

$t^2x'' - tx' + x = t^2 \cdot 0 - t \cdot 1 + t = 0$, so t is a solution.

$x = tz \Longrightarrow x' = tz' + z \Longrightarrow x'' = tz'' + 2z'$.

Thus,

$t^2x'' - tx' + x = t^2(tz'' + 2z') - t(tz' + z) + tz = 0$

reduces to

$t^2(tz'' + z') = 0$, or $tz'' + z' = 0$ because $t > 0$.

Put $v = z'$ to find the separable equation $tv' + v = 0$ with solution $v = C_1/t$.

Solving $z' = C_1/t$ gives $z = C_1 \ln t + C_2$, so

$x(t) = t(C_1 \ln t + C_2)$ is the solution of $t^2x'' - tx' + x = 0$.

This solution has at most one zero if $t > 0$. If we try to apply the Oscillation and Relation Theorems to $x'' - x'/t + x/t^2 = 0$, we find $Q(t) = 1/t^2 - (1/2)/t^2 - (1/4)/t^2 = (1/4)/t^2$. Although this is positive, we cannot find a constant k for which $Q(t) \geq k^2$, so the theorems do not apply.

(e) $x = t\sin t \Longrightarrow x' = \sin t + t\cos t \Longrightarrow x'' = 2\cos t - t\sin t$.

$t^2x'' - 2tx' + (t^2+2)\,x$

$= t^2(2\cos t - t\sin t) - 2t(\sin t + t\cos t) + (t^2+2)\,t\sin t$

$= 0$, so $t\sin t$ is a solution.

$x = zt\sin t \Longrightarrow$

$x' = z(\sin t + t\cos t) + z't\sin t \Longrightarrow$

$x'' = z(2\cos t - t\sin t) + 2z'(\sin t + t\cos t) + z''t\sin t$.

Thus, $t^2x'' - 2tx' + (t^2+2)\,x$

$= t^2[z(2\cos t - t\sin t) + 2z'(\sin t + t\cos t) + z''t\sin t] +$

$\quad -2t[z(\sin t + t\cos t) + z't\sin t] + (t^2+2)\,zt\sin t$

$= z''t^3\sin t + z'\left[2t^2(\sin t + t\cos t) - 2t^2\sin t\right]$

$= z''t^3\sin t + 2z't^3\cos t$

$= 0$,

or $z''\sin t + 2z'\cos t = 0$.

Put $v = z'$ to find the separable equation $v'\sin t + 2v\cos t = 0$ with solution $v = C_1/\sin^2 t$.

Solving $z' = C_1/\sin^2 t$ gives $z = -C_1\cos t/\sin t + C_2$, so

$x(t) = t\sin t(-C_1\cos t/\sin t + C_2) = -C_1 t\cos t + C_2 t\sin t$ is the solution of $t^2x'' - 2tx' + (t^2+2)\,x = 0$.

This solution oscillates.

If we apply the Oscillation and Relation Theorems to $x'' - 2x'/t + (1 + 2/t^2)\,x = 0$, we find

$Q(t) = (1 + 2/t^2) - (1/2)(-2/t)' - (1/4)(-2/t)^2 = 1 \geq 1$,

which requires an infinite number of zeros.

(g) $x = t \Longrightarrow x' = 1 \Longrightarrow x'' = 0 \Longrightarrow$

$t^2x'' - t(t+2)\,x' + (t+2)\,x = t^2 \cdot 0 - t(t+2) \cdot 1 + (t+2) \cdot t = 0$, so t is a solution.

$x = tz \Longrightarrow x' = tz' + z \Longrightarrow x'' = tz'' + 2z'$.
Thus,
$t^2 x'' - t(t+2)x' + (t+2)x$
$= t^2(tz'' + 2z') - t(t+2)(tz' + z) + (t+2)tz$
$= 0$
reduces to $t^3(z'' - z') = 0$, or $z'' - z' = 0$.
Although we could put $v = z'$ to find a separable equation, we can write down the solution of $z'' - z' = 0$ immediately as $z = C_1 + C_2 e^t$, so
$x(t) = t(C_1 + C_2 e^t)$ is the solution of $t^2 x'' - t(t+2)x' + (t+2)x = 0$.
This solution has at most one zero if $t > 0$.

3. **(a)** The general solution of $x'' + 2x' + x = 0$ is $x(t) = (C_1 + C_2 t)e^{-t}$, so $x_1(t) = e^{-t}$ is a solution.
Let $x = ze^{-t}$, so that $x' = (z' - z)e^{-t}$, $x'' = (z'' - 2z' + z)e^{-t}$, and
$x'' + 2x' + x = z'' e^{-t}$. Thus,
$z'' e^{-t} = 2e^{-t}/t^2$ can be integrated to yield
$z' = -\frac{2}{t} + C_1$, and
$z = -2\ln t + C_1 t + C_2$.
Thus,
$x(t) = ze^{-t} = (-2\ln t + C_1 t + C_2)e^{-t}$.

(c) The general solution of $x'' + x = 0$ is $x(t) = C_1 \cos t + C_2 \sin t$, so $x_1(t) = \cos t$ is a solution.
Let $x = z\cos t$, so that $x' = z'\cos t - z\sin t$, $x'' = z''\cos t - 2z'\sin t - z\cos t$, and
$x'' + x = z''\cos t - 2z'\sin t$. Thus,
$z''\cos t - 2z'\sin t = \sec t$. This is a linear differential equation in z', which when put in standard form, has an integrating factor $\mu = \cos^2 t$. This gives
$z''\cos^2 t - 2z'\cos t \sin t = 1$ which can be integrated to yield
$z'\cos^2 t = t + C_1$, or
$z' = t\sec^2 t + C_1 \sec^2 t$. Thus,
$z = \int t\sec^2 t\,dt + C_1 \int \sec^2 t\,dt + C_2$. Using integration by parts we find
$\int t\sec^2 t\,dt = t\tan t - \int \tan t\,dt = t\tan t + \ln\cos t$, so
$z = t\tan t + \ln\cos t + C_1 \tan t + C_2$
Thus,
$x(t) = z\cos t$
$= (t\tan t + \ln\cos t + C_1\tan t + C_2)\cos t$
$= t\sin t + \cos t \ln\cos t + C_1 \sin t + C_2 \cos t$.

5. **(a)** Here $a_2 = 1$, $a_1 = 2$, $a_0 = 1$, $f(t) = 2e^{-t}/t^2$, and $u(t) = e^{-t}$, so
$\alpha(t) = \exp\left(\int 2\,dt\right) = e^{2t}$ and

$$x(t) = \int \frac{1}{e^{-2t}e^{2t}} \left(\int \frac{e^{-t}e^{2t}2e^{-t}}{t^2}\,dt \right) dt = \int \left(\int \frac{2}{t^2}\,dt \right) dt = \int -\frac{2}{t}\,dt = -2\ln t.$$

(c) Here $a_2 = 1$, $a_1 = 0$, $a_0 = 1$, $f(t) = \sec t$, and $u(t) = \cos t$, so
$\alpha(t) = \exp\left(\int 0\,dt\right) = e^0 = 1$ and

$$x(t) = \int \frac{1}{\cos^2 t} \left(\int \frac{\cos t \sec t}{1}\,dt \right) dt = \int t\sec^2 t\,dt.$$

Using integration by parts we find
$\int t\sec^2 t\,dt = t\tan t - \int \tan t\,dt = t\tan t + \ln\cos t$, so
$x(t) = t\tan t - \int \tan t\,dt = t\tan t + \ln\cos t$.

7. **(a)** $x = e^t$, $x' = e^t$, $x'' = e^t$, so that
$(t^2+t)\,x'' + (2-t^2)\,x' - (2+t)\,x = (t^2+t)\,e^t + (2-t^2)\,e^t - (2+t)\,e^t = 0.$

(b) $x_p = e^t z$, so that $x_p' = e^t z' + e^t z$ and $x_p'' = e^t z'' + 2e^t z' + e^t z$ and
$(t^2+t)\,(e^t z'' + 2e^t z' + e^t z) + (2-t^2)\,(e^t z' + e^t z) - (2+t)\,e^t z = (t+1)^2$, or
$(t^2+t)\,z'' + (t^2+2t+2)\,z' = (t+1)^2\,e^{-t}$, or

$$z'' + \left[1 + \frac{t+2}{t\,(t+1)}\right] z' = \frac{t+1}{t}e^{-t}.$$

This is a linear differential equation for z' with integrating factor

$$\mu = \exp\int\left[1+\frac{t+2}{t\,(t+1)}\right]dt = \exp\left[t + \ln\left(\frac{t^2}{t+1}\right)\right] = \frac{t^2}{t+1}e^t$$

giving

$$\frac{t^2}{t+1}e^t z'' + \frac{t^2}{t+1}e^t\left[1+\frac{t+2}{t\,(t+1)}\right]z' = t$$

with solution

$$\frac{t^2}{t+1}e^t z' = \frac{1}{2}t^2 + C_1,$$

so

$$z' = \frac{1}{2}(t+1)\,e^{-t} + C_1\frac{t+1}{t^2}e^{-t}.$$

Integration gives

$$z = \frac{1}{2}\left(-te^{-t} - 2e^{-t}\right) - C_1\frac{1}{t}e^{-t} + C_2$$

so

$$x\,(t) = e^t z = \frac{1}{2}(-t-2) - C_1\frac{1}{t} + C_2 e^t.$$

8.3 Finding Solutions by Variation of Parameters

1. If $x_1 = cx_2\,(t)$ then $W\,[x_1, x_2] = x_1x_2' - x_1'x_2 = cx_2x_2' - cx_2'x_2 = 0.$

3. **(a)** The general solution of the associated homogeneous equation is
$x_h(t) = C_1e^{-t} + C_2te^{-t}.$
We assume a particular solution of the form $x_p(t) = z_1e^{-t} + z_2te^{-t}$, so that
$z_1'e^{-t} + z_2'te^{-t} = 0$ and
$-z_1'e^{-t} + z_2'\,(-t+1)\,e^{-t} = 2t^{-2}e^{-t}.$
Thus,
$z_1' + z_2't = 0$ and
$-z_1' + z_2'\,(-t+1) = 2t^{-2}$, which, when added, give $z_2' = 2t^{-2}$.
Integration gives $z_2 = -2t^{-1} + C_2$.
From $z_1' + z_2't = 0$ we find $z_1' = -2t^{-1}$, so $z_1 = -2\ln t + C_1$. Thus,
$x(t) = (-2\ln t + C_1)\,e^{-t} + \left(-2t^{-1} + C_2\right)te^{-t}$
$= -2e^{-t}\ln t - 2e^{-t} + C_1e^{-t} + C_2te^{-t}.$

(c) The general solution of the associated homogeneous equation is
$x_h(t) = C_1 \cos t + C_2 \sin t$.
We assume a particular solution of the form $x_p(t) = z_1 \cos t + z_2 \sin t$, so that
$z_1' \cos t + z_2' \sin t = 0$ and
$-z_1' \sin t + z_2' \cos t = \sec t$.
Multipying the first equation by $\sin t$ and the second by $\cos t$ and adding gives
$z_2' = 1$ and $z_2 = t + C_2$.
Also, $z_1' = -\frac{\sin t}{\cos t}$ so $z_1 = \ln \cos t + C_1$.
Thus, $x(t) = (\ln \cos t + C_1)\cos t + (t + C_2)\sin t$.

5. The solution of the associated homogeneous equation is
$x_h(t) = C_1 e^{-t} + C_2 e^t$.
We assume a particular solution of the form $x_p(t) = z_1 e^{-t} + z_2 e^t$, so that
$z_1' e^{-t} + z_2' e^t = 0$ and
$-z_1' e^{-t} + z_2' e^t = 1/t$.
Thus, $2z_2' e^t = 1/t$ and $z_2' = \frac{1}{2} e^{-t}/t$ so $z_2 = \int_1^t \frac{1}{2} e^{-s}/s\, ds + C_2$.
Also, $z_1' = -e^{2t} z_2' = -\frac{1}{2} e^t/t$. Integrating this gives $z_1 = -\int_1^t \frac{1}{2} e^s/s\, ds + C_1$.
Thus, $x(t) = e^{-t}\left(-\int_1^t \frac{1}{2} e^s/s\, ds + C_1\right) + e^t\left(\int_1^t \frac{1}{2} e^{-s}/s\, ds + C_2\right)$.
Now,
$x'(t) = -e^{-t}\left(-\int_1^t \frac{1}{2} e^s/s\, ds + C_1\right) + e^{-t}\left(-\frac{1}{2} e^t/t\right) + e^t\left(\int_1^t \frac{1}{2} e^{-s}/s\, ds + C_2\right) + e^t\left(\frac{1}{2} e^{-t}/t\right)$,
so $A = x(1) = e^{-1} C_1 + e C_2$ and $B = x'(1) = -e^{-1} C_1 + e C_2$.
Solving for C_1 and C_2 gives $C_2 = e^{-1}(A + B)/2$ and $C_1 = e(A - B)/2$.
This gives
$x(t) = e^{-t}\left[-\int_1^t \frac{1}{2} e^s/s\, ds + e(A - B)/2\right] + e^t\left[\int_1^t \frac{1}{2} e^{-s}/s\, ds + e^{-1}(A + B)/2\right]$
$= \frac{1}{2}\left[e^{t-1}(A + B) + e^{-t+1}(A - B) + e^t \int_1^t e^{-s}/s\, ds - \int_1^t e^s/s\, ds\right]$.

8.4 The Importance of Linear Independence and Dependence

1. (a) $\begin{vmatrix} t & t^2 \\ 1 & 2t \end{vmatrix} = t^2$, which is not identically zero.

(c) $\begin{vmatrix} 1 & e^t \\ 0 & e^t \end{vmatrix} = e^t$, which is not identically zero.

(e) $\begin{vmatrix} \sinh 2t & \cosh 2t \\ 2\cosh 2t & 2\sinh 2t \end{vmatrix}$
$= 2\sinh^2 2t - 2\cosh^2 2t$
$= -2$, which is not identically zero.

(g) $\begin{vmatrix} e^{-t} & te^{-t} \\ -e^{-t} & e^{-t} - te^{-t} \end{vmatrix} = e^{-2t}$, which is not identically zero.

(i) $\begin{vmatrix} e^{\alpha t}\cos\beta t & e^{\alpha t}\sin\beta t \\ \alpha e^{\alpha t}\cos\beta t - \beta e^{\alpha t}\sin\beta t & \alpha e^{\alpha t}\sin\beta t + \beta e^{\alpha t}\cos\beta t \end{vmatrix}$
$= e^{\alpha t}\cos\beta t\,(\alpha e^{\alpha t}\sin\beta t + \beta e^{\alpha t}\cos\beta t) - e^{\alpha t}\sin\beta t\,(\alpha e^{\alpha t}\cos\beta t - \beta e^{\alpha t}\sin\beta t)$
$= \beta e^{\alpha t}\left(\cos^2\beta t + \sin^2\beta t\right) = \beta e^{\alpha t}$, which is not identically zero.

(k) $\begin{vmatrix} t^a \cos(b\ln t) & t^a \sin(b\ln t) \\ at^{a-1}\cos(b\ln t) - bt^{a-1}\sin(b\ln t) & at^{a-1}\sin(b\ln t) + bt^{a-1}\cos(b\ln t) \end{vmatrix}$
$= bt^{2a-1}\left[\cos^2(b\ln t) + \sin^2(b\ln t)\right]$
$= bt^{2a-1}$, which is not identically zero.

3. (a) $\begin{vmatrix} t & k+t \\ 1 & 1 \end{vmatrix} = -k$, which is not identically zero if $k \neq 0$.

(c) $\begin{vmatrix} 1+2t & k+t \\ 2 & 1 \end{vmatrix} = 1-2k$, which is not identically zero if $k \neq \frac{1}{2}$.

(e) $\begin{vmatrix} 3 & 2+kt \\ 0 & k \end{vmatrix} = 3k$, which is not identically zero if $k \neq 0$.

(g) $\begin{vmatrix} t^2 & (t+k)^2 \\ 2t & 2(t+k) \end{vmatrix}$
$= 2t^2(t+k) - 2t(t+k)^2$
$= -2kt(t+k)$, which is not identically zero if $k \neq 0$.

(i) $\sin^2 kt + \cos^2 kt = 1 = \frac{1}{6}(6)$ so they are linearly dependent for all values of k.

(k) $\begin{vmatrix} e^{kt} & te^{kt} \\ ke^{kt} & (1+kt)e^{kt} \end{vmatrix}$
$= (1+kt)e^{2kt} - kte^{2kt}$
$= e^{2kt}$, which is not identically zero for all k.

5. (a) $a_2x_1'' + a_1x_1' + a_0x_1 = 0$ and $a_2x_2'' + a_1x_2' + a_0x_2 = 0$. If we multiply the first equation by x_2 and the second by x_1 and then subtract the resulting equations, we find

$$x_2(a_2x_1'' + a_1x_1' + a_0x_1) - x_1(a_2x_2'' + a_1x_2' + a_0x_2) = 0,$$

or

$$a_2(x_1''x_2 - x_2''x_1) + a_1(x_1'x_2 - x_2'x_1) = 0,$$

which is the required equation.

(b) From $W = x_1'x_2 - x_2'x_1$ we have $W' = x_1''x_2 - x_2''x_1$ which when substituted in the equation from part (a) gives $a_2W' + a_1W = 0$. This is a separable equation with solution $W = C\exp\left(-\int a_1/a_2\,dt\right)$, where C is a constant. (Note: $C = 0$ gives the equilibrium solution, and $C \neq 0$ gives the nonequilibrium solutions.)

(c) We can write $W(t) = Cf(t)$ where $f(t) = \exp\left(-\int a_1/a_2\,dt\right)$ and $f(t) > 0$. If $W = 0$ at any point $t = t_0$, that is, $W(t_0) = 0$, then $Cf(t_0) = 0$, and so $C = 0$, in which case $W(t) = 0$ for all t. Thus, if $W \neq 0$ at any point $t = t_0$, then $W(t) \neq 0$ for all t.

(d) If x_1 and x_2 are linearly independent, then there must be at least one point at which $W \neq 0$, and part (c) then guarantees that $W(t) \neq 0$ for all t.

If x_1 and x_2 are linearly dependent, then there must be at least one point at which $W = 0$, and part (c) then guarantees that $W(t) = 0$ for all t.

7. Let the upper curve be $x_1(t)$ and the lower $x_2(t)$. It appears that $x_1(t)$ has a maximum at the same place that $x_2(t)$ has a minimum, namely, $t = -4$. Thus, $x_1'(-4) = x_2'(-4) = 0$ so $W(-4) = x_1'x_2 - x_2'x_1 = 0$. This guarantees that the solutions are linearly dependent.

9. The general solution of $a_2x'' + a_1x' + a_0x = f(t)$ is $x(t) = x_h(t) + x_p(t)$, where $x_h(t)$ is the general solution of $a_2x_h'' + a_1x_h' + a_0x_h = 0$, and $x_p(t)$ is any solution of $a_2x_p'' + a_1x_p' + a_0x_p = f(t)$. According to Exercise 8, $x_h = x_1 + x_2$, while x_p is the particular solution that satisfies $a_2x_p'' + a_1x_p' + a_0x_p = f(t)$, $x(0) = 0$, $x'(0) = 0$. Also, $x(0) = x_1(0) + x_2(0) + x_p(0) = x_0$ and $x'(0) = x_1'(0) + x_2'(0) + x_p'(0) = x_0^*$.

11. **(a)** $\begin{vmatrix} t & t^2 & x \\ 1 & 2t & x' \\ 0 & 2 & x'' \end{vmatrix}$

$= x''\left(2t^2 - t^2\right) - x'(2t - 0) + x(2 - 0)$

$= t^2x'' - 2tx' + 2x$. Thus,

$t^2x'' - 2tx' + 2x = 0$ has t and t^2 as solutions.

(c) $\begin{vmatrix} t & e^t & x \\ 1 & e^t & x' \\ 0 & e^t & x'' \end{vmatrix}$

$= x''\left(te^t - e^t\right) - x'\left(te^t - 0\right) + x\left(e^t - 0\right)$

$= e^t\left[(t-1)x'' - tx' + x\right]$. Thus,

$(t-1)x'' - tx' + x = 0$ has t and e^t as solutions.

(e) $\begin{vmatrix} e^{-t} & te^{-t} & x \\ -e^{-t} & e^{-t} - te^{-t} & x' \\ e^{-t} & -2e^{-t} + te^{-t} & x'' \end{vmatrix} = e^{-2t}\begin{vmatrix} 1 & t & x \\ -1 & 1-t & x' \\ 1 & -2+t & x'' \end{vmatrix}$

$= x'' + 2x' + x$. Thus,

$x'' + 2x' + x = 0$ has e^{-t} and te^{-t} as solutions.

(g) $\begin{vmatrix} \sin\beta t & \cos\beta t & x \\ \beta\cos\beta t & -\beta\sin\beta t & x' \\ -\beta^2\sin\beta t & -\beta^2\cos\beta t & x'' \end{vmatrix}$

$= x''\left(-\beta\sin^2\beta t - \beta\cos^2\beta t\right) - x'\left(-\beta^2\cos\beta t\sin\beta t + \beta^2\cos\beta t\sin\beta t\right) +$

$\quad + x\left(-\beta^3\cos^2\beta t - \beta^3\sin^2\beta t\right)$

$= -\beta x'' - \beta^3 x$. Thus,

$x'' + \beta^2 x = 0$ has $\sin\beta t$ and $\cos\beta t$ as solutions.

(i) $\begin{vmatrix} \sin t & \sin^2 t & x \\ \cos t & 2\sin t\cos t & x' \\ -\sin t & 2\cos^2 t - 2\sin^2 t & x'' \end{vmatrix}$

$= x''\left(\sin^2 t\cos t\right) - x'\left(2\sin t\cos^2 t - \sin^3 t\right) + x\left(2\cos^3 t - 2\cos t\sin^2 t + 2\sin^2 t\cos t\right)$

$= x''\left(\sin^2 t\cos t\right) - x'\left(2\sin t\cos^2 t - \sin^3 t\right) + x\left(2\cos t - 2\cos t\sin^2 t\right)$. Thus,

$x''\left(\sin^2 t\cos t\right) - x'\left(2\sin t\cos^2 t - \sin^3 t\right) + x\left(2\cos t - 2\cos t\sin^2 t\right) = 0$ has $\sin t$ and $\sin^2 t$ as solutions.

(k) $\begin{vmatrix} e^t & \sin t & x \\ e^t & \cos t & x' \\ e^t & -\sin t & x'' \end{vmatrix}$

$= e^t\begin{vmatrix} 1 & \sin t & x \\ 1 & \cos t & x' \\ 1 & -\sin t & x'' \end{vmatrix}$

$= e^t\left[x''(\cos t - \sin t) - x'(-\sin t - \sin t) + x(-\sin t - \cos t)\right]$

$= e^t\left[(\cos t - \sin t)x'' + 2(\sin t)x' - (\sin t + \cos t)x\right]$. Thus,

$(\cos t - \sin t)x'' + 2(\sin t)x' - (\sin t + \cos t)x = 0$ has e^t and $\sin t$ as solutions.

13. With $x_1 = w$ and $x_2 = w^{-1}$, we have

$x_1' = w'$, $x_1'' = w''$, $x_2' = -w^{-2}w'$, $x_2'' = 2w^{-3}(w')^2 - w^{-2}w''$.

Thus, $x_1x_2' - x_1'x_2 = w\left(-w^{-2}w'\right) - w'\left(w^{-1}\right) = -2w^{-1}w'$.

$x_1x_2'' - x_1''x_2 = w\left(2w^{-3}(w')^2 - w^{-2}w''\right) - w''\left(w^{-1}\right) = 2w^{-2}(w')^2 - 2w^{-1}w'' = -2w^{-1}w'\left(-\frac{w'}{w} + \frac{w''}{w'}\right)$.

$x_1'x_2'' - x_1''x_2' = w'\left(2w^{-3}(w')^2 - w^{-2}w''\right) - w''\left(-w^{-2}w'\right) = 2w^{-3}(w')^3 = 2w^{-1}w'\left(\frac{w'}{w}\right)^2$.

Thus,

$p = -\frac{x_1 x_2'' - x_1'' x_2}{x_1 x_2' - x_1' x_2} = -\left(-2w^{-1}w'\right)\left(-\frac{w'}{w} + \frac{w''}{w'}\right) / \left(-2w^{-1}w'\right) = \frac{w'}{w} - \frac{w''}{w'}$, and

$q = \frac{x_1' x_2'' - x_1'' x_2'}{x_1 x_2' - x_1' x_2} = 2w^{-1}w'\left(\frac{w'}{w}\right)^2 / \left(-2w^{-1}w'\right) = -\left(\frac{w'}{w}\right)^2.$

15. **(a)** $x_1(t) = 1$ gives $x_1' = 0$ so $(\sin t \cos t)\, x' - x + x^3 = 0$.

$x_1(t) = \sin t$ gives $x_1' = \cos t$ so $(\sin t \cos t)\, x' - x + x^3$
$= (\sin t \cos t)\cos t - \sin t + \sin^3 t$
$= \sin t\left(\cos^2 t + \sin^2 t\right) - \sin t = 0.$

(b) $\begin{vmatrix} 1 & \sin t \\ 0 & \cos t \end{vmatrix} = \cos t$, which is not identically zero.

(c) $x(t) = C_1 + C_2 \sin t$ gives $x' = C_2 \cos t$ so $(\sin t \cos t)\, x' - x + x^3$
$= (\sin t \cos t)\, C_2 \cos t - (C_1 + C_2 \sin t) + (C_1 + C_2 \sin t)^3$
$= C_2 \sin t \cos^2 t - C_1 - C_2 \sin t + C_1^3 + 3C_1^2 C_2 \sin t + 3C_1 C_2^2 \sin^2 t + C_2^3 \sin^3 t$
$\neq 0$. It is not a solution.

8.5 Solving Cauchy-Euler Equations

1. **(a)** $x^2 \frac{d^2y}{dx^2} - 2y = 0 \Longrightarrow \frac{d^2y}{dt^2} - \frac{dy}{dt} - 2y = 0$ when $x = e^t$.
Characteristic equation $r^2 - r - 2 = 0$.
Roots $r_1 = -1$, $r_2 = 2$.
Solution $y = C_1 e^{-t} + C_2 e^{2t}$.
Because $x = e^t$, $y(x) = C_1/x + C_2 x^2$.

(c) $x^2 \frac{d^2y}{dx^2} + x\frac{dy}{dx} = 0 \Longrightarrow \frac{d^2y}{dt^2} - \frac{dy}{dt} + \frac{dy}{dt} = 0 \Longrightarrow \frac{d^2y}{dt^2} = 0$ when $x = e^t$.
Characteristic equation $r^2 = 0$.
Repeated root $r = 0$.
Solution $y = C_1 + C_2 t$.
Because $x = e^t$, $y(x) = C_1 + C_2 \ln x$.

(e) $x^2 \frac{d^2y}{dx^2} + x\frac{dy}{dx} + 9y = 0 \Longrightarrow \frac{d^2y}{dt^2} - \frac{dy}{dt} + \frac{dy}{dt} + 9y = 0 \Longrightarrow \frac{d^2y}{dt^2} + 9y = 0$ when $x = e^t$.
Characteristic equation $r^2 + 9 = 0$.
Roots $r_{1,2} = \pm 3i$.
Solution $y = C_1 \cos 3t + C_2 \sin 3t$.
Because $x = e^t$, $y(x) = C_1 \cos(3 \ln x) + C_2 \sin(3 \ln x)$.

(g) $x^2 \frac{d^2y}{dx^2} + 5x\frac{dy}{dx} + 4y = 0 \Longrightarrow \frac{d^2y}{dt^2} - \frac{dy}{dt} + 5\frac{dy}{dt} + 4y = 0 \Longrightarrow \frac{d^2y}{dt^2} + 4\frac{dy}{dt} + 4y = 0$ when $x = e^t$.
Characteristic equation $r^2 + 4r + 4 = 0$.
Repeated root $r = -2$.
Solution $y = C_1 e^{-2t} + C_2 t e^{-2t}$.
Because $x = e^t$, $y(x) = \frac{C_1}{x^2} + \frac{C_2}{x^2} \ln x$.

(i) $x^2 \frac{d^2y}{dx^2} - 5x\frac{dy}{dx} + 5y = 0 \Longrightarrow \frac{d^2y}{dt^2} - \frac{dy}{dt} - 5\frac{dy}{dt} + 5y = 0 \Longrightarrow \frac{d^2y}{dt^2} - 6\frac{dy}{dt} + 5y = 0$ when $x = e^t$.
Characteristic equation $r^2 - 6r + 5 = 0$.
Roots $r_1 = 1$, $r_2 = 5$.
Solution $y = C_1 e^t + C_2 e^{5t}$.
Because $x = e^t$, $y(x) = C_1 x + C_2 x^5$.

(k) $x^2\frac{d^2y}{dx^2} - x\frac{dy}{dx} + y = 0 \Longrightarrow \frac{d^2y}{dt^2} - \frac{dy}{dt} - \frac{dy}{dt} + y = 0 \Longrightarrow \frac{d^2y}{dt^2} - 2\frac{dy}{dt} + y = 0$ when $x = e^t$.
Characteristic equation $r^2 - 2r + 1 = 0$.
Repeated root $r = 1$.
Solution $y = C_1e^t + C_2te^t$.
Because $x = e^t$, $y(x) = C_1x + C_2x\ln x$.

3. (a) $x^2\frac{d^2y}{dx^2} - 2y = \ln x \Longrightarrow \frac{d^2y}{dt^2} - \frac{dy}{dt} - 2y = t$ when $x = e^t$.
Characteristic equation $r^2 - r - 2 = 0$.
Roots $r_{1,2} = -1, 2$.
Solution $y_h = C_1e^{-t} + C_2e^{2t}$.
$y_p = At + B,\ \frac{dy_p}{dt} = A,\ \frac{d^2y_p}{dt^2} = 0$
$\frac{d^2y}{dt^2} - \frac{dy}{dt} - 2y = 0 - A - 2(At + B)$
$= -2At - 2B - A$
$= t$
so $A = -1/2$, $B = -A/2 = 1/4$, and $y_p = -\frac{1}{2}\left(t - \frac{1}{2}\right)$.
Thus, $y = y_p + y_h = -\frac{1}{2}\left(t - \frac{1}{2}\right) + C_1e^{-t} + C_2e^{2t}$.
Because $x = e^t$, $y(x) = -\frac{1}{2}\left(\ln x - \frac{1}{2}\right) + C_1/x + C_2x^2$.

(c) $x^2\frac{d^2y}{dx^2} + 7x\frac{dy}{dx} + 5y = 10 - 4x^{-1} \Longrightarrow \frac{d^2y}{dt^2} - \frac{dy}{dt} + 7\frac{dy}{dt} + 5y = 10 - 4e^{-t} \Longrightarrow$
$\frac{d^2y}{dt^2} + 6\frac{dy}{dt} + 5y = 10 - 4e^{-t}$ when $x = e^t$.
Characteristic equation $r^2 + 6r + 5 = 0$.
Roots $r_{1,2} = -1, -5$.
Solution $y_h = C_1e^{-t} + C_2e^{-5t}$.
$y_p = A + Bte^{-t},\ \frac{dy_p}{dt} = -Bte^{-t} + Be^{-t},\ \frac{d^2y_p}{dt^2} = Bte^{-t} - 2Be^{-t}$
$\frac{d^2y_p}{dt^2} + 6\frac{dy_p}{dt} + 5y_p = Bte^{-t} - 2Be^{-t} + 6(-Bte^{-t} + Be^{-t}) + 5(A + Bte^{-t})$
$= 4Be^{-t} + 5A$
$= 10 - 4e^{-t}$
so $B = -1$, $A = 2$, and $y_p = 2 - te^{-t}$.
Thus, $y = y_p + y_h = 2 - te^{-t} + C_1e^{-t} + C_2e^{-5t}$.
Because $x = e^t$, $y(x) = 2 - \ln x/x + C_1/x + C_2/x^5$.

5. $\frac{d^2y}{dx^2} - \frac{a^2}{x^2}y = 0 \Longrightarrow x^2\frac{d^2y}{dx^2} - a^2y = 0 \Longrightarrow \frac{d^2y}{dt^2} - \frac{dy}{dt} - a^2y = 0$ when $x = e^t$.
Characteristic equation $r^2 - r + a^2 = 0$.
Roots $r_{1,2} = \left(1 \pm \sqrt{1 - 4a^2}\right)/2$.

(a) If $a^2 < 1/4$ then the roots are real and distinct.
Solution $y = C_1e^{r_1t} + C_2e^{r_2t}$.
Because $x = e^t$, $y(x) = C_1x^{r_1} + C_2x^{r_2}$.

(b) If $a^2 = 1/4$ then the root is repeated and $r = 1/2$.
Solution $y = C_1e^{t/2} + C_2te^{t/2}$.
Because $x = e^t$, $y(x) = C_1\sqrt{x} + C_2\sqrt{x}\ln x$.

(c) If $a^2 > 1/4$ then the roots are complex, namely $r_{1,2} = 1/2 \pm \omega i$ where $\omega = \left(\sqrt{4a^2 - 1}\right)/2$.
Solution $y = C_1e^{t/2}\cos\omega t + C_2e^{t/2}\sin\omega t$.
Because $x = e^t$, $y(x) = C_1\sqrt{x}\cos(\omega\ln x) + C_2\sqrt{x}\sin(\omega\ln x)$.

7. If $y = x^r$ then $y' = rx^{r-1}$ and $y'' = r(r-1)x^{r-2}$. Thus,

$$\begin{aligned} &b_2x^2y'' + b_1xy' + b_0y \\ &= b_2x^2r(r-1)x^{r-2} + b_1xrx^{r-1} + b_0x^r \\ &= [b_2r(r-1) + b_1r + b_0]x^r \\ &= 0, \text{ gives} \\ &b_2r(r-1) + b_1r + b_0 \\ &= b_2r^2 + (b_1 - b_2)r + b_0 \\ &= 0 \text{ with roots} \\ &r_{1,2} = \left[-(b_1 - b_2) \pm \sqrt{(b_1 - b_2)^2 - 4b_2b_0}\right] / [2b_2]. \end{aligned}$$

The nature of the roots will depend on the sign of $(b_1 - b_2)^2 - 4b_2b_0$.

If $(b_1 - b_2)^2 - 4b_2b_0 > 0$ then the roots will be real and distinct and the general solution will be $y(x) = C_1x^{r_1} + C_2x^{r_2}$.

If $(b_1 - b_2)^2 - 4b_2b_0 = 0$ then the root will be repeated so $r_1 = r_2 = r$. One solution will be $y(x) = x^r$. The other solution can be obtained by reduction of order using $y = x^r z$. The general solution will be $y(x) = C_1x^r + C_2x^r \ln x$.

If $(b_1 - b_2)^2 - 4b_2b_0 < 0$ then the roots will be $r_{1,2} = \alpha \pm \beta i$ where α, β are real. The solution could be written $y(x) = C_1x^{\alpha-\beta i} + C_2x^{\alpha+\beta i}$. This can be rewritten, using $x^{a+bi} = x^a x^{bi} = x^a \exp(ib \ln x) = x^a [\cos(b \ln x) + i \sin(b \ln x)]$, as
$y(x) = C_1x^{\alpha t}[\cos(-\beta \ln x) + i \sin(-\beta \ln x)] + C_2x^{\alpha t}[\cos(\beta \ln x) + i \sin(\beta \ln x)]$ or
$y(x) = K_1x^{\alpha t}\cos(\beta \ln x) + K_2x^{\alpha t}\sin(\beta \ln x)$ where $K_1 = C_1 + C_2$ and $K_2 = i(-C_1 + C_2)$.

8.6 Boundary Value Problems and the Shooting Method

1. (a) By the superposition principle, the linear combination, $x(t) = \alpha x_1(t) + (1-\alpha)x_2(t)$, of x_1 and x_2 satisfies $a_2x'' + a_1x' + a_0x = 0$, because x_1 and x_2 do. Thus, the only thing left to check is that x satisfies the boundary conditions $x(a) = x_a$, $x(b) = x_b$. Now, $x(a) = \alpha x_1(a) + (1-\alpha)x_2(a) = \alpha x_a + (1-\alpha)x_a = x_a$. Similarly, $x(b) = x_b$.

   (b) $a_2(\alpha x_2)'' + a_1(\alpha x_2)' + a_0(\alpha x_2) = \alpha(a_2x_2'' + a_1x_2' + a_0x_2) = 0$, while $\alpha x_2(a) = 0$.

   (c) Assume that the boundary value problem has two distinct solutions. There are two possibilities, either both solutions are nontrivial, or one of them is the trivial solution. In the first case from the two nontrivial solutions we are able to construct an infinite number of solutions, namely, from part (a), $x(t) = \alpha x_1(t) + (1-\alpha)x_2(t)$. In the second case, because the trivial solution is a solution, the boundary conditions must be $x(a) = 0$ and $x(b) = 0$. From part (b), $x(t) = \alpha x_2(t)$ then gives us an infinite number of solutions.

3. (a) The general solution of $x'' + 100x = 0$ is $x(t) = C_1 \cos 10t + C_2 \sin 10t$.
   $x(0) = 4$ requires $4 = C_1$, while $x(\pi/4) = 500$ requires that
   $500 = C_1 \cos \frac{5}{2}\pi + C_2 \sin \frac{5}{2}\pi = C_2$,
   $x(t) = 4\cos 10t + 500 \sin 10t$.

(c) The general solution of $x'' + 2x' + 5x = 0$ is

$x(t) = e^{-t}(C_1 \cos 2t + C_2 \sin 2t)$.

$x(0) = 0$ requires $0 = C_1$, while $x(\pi/4) = 0$ requires that $0 = C_2$,

$x(t) = 0$.

5. $x'' + 16x = 0$ has solution $x(t) = C_1 \cos 4t + C_2 \sin 4t$, so $x(0) = 0$ gives $C_1 = 0$, and $x(b) = 0$ then gives $C_2 \sin 4b = 0$.

(a) This is not possible, because the trivial solution is always a solution in this case.

(b) Because the trivial solution is a solution we would want $C_2 = 0$ to be the only consequence of $C_2 \sin 4b = 0$. Thus, we want $\sin 4b \neq 0$, that is, $b \neq n\pi/4$, $n = 0, \pm 1, \pm 2, \cdots$.

(c) To have an infinite number of solutions we would want $\sin 4b = 0$, that is, $b = n\pi/4$, $n = 0, \pm 1, \pm 2, \cdots$. In this case the solutions are $x(t) = C_2 \sin 4t$, for any C_2.

7. $x'' + \lambda^2 x = 0$ has the solution $x(t) = C_1 \cos \lambda t + C_2 \sin \lambda t$. $x(0) = 0$ gives $C_1 = 0$, so $x(t) = C_2 \sin \lambda t$, and $x'(t) = \lambda C_2 \cos \lambda t$. $x'(\pi) = 0$ gives $\lambda C_2 \cos \lambda \pi = 0$, so $\lambda = (2n+1)/2$, $n = 0, \pm 1, \pm 2, \cdots$. Thus,

$x(t) = C_2 \sin[(2n+1)t/2]$.

9. The solution of the second boundary value problem is not unique, because the first boundary value problem has a nontrivial solution. See the first comment after The Existence Theorem for Boundary Value Problems.

11. With $x_0^* \approx 1.1$ the solution passes through the point $(2, 1)$.

With $x_0^* = 0$ the solution passes through the point $(2, 0)$.

There are no other solutions.

8.7 Solving Higher Order Homogeneous Differential Equations

1. (a) The characteristic equation $r^3 + r = 0$ has roots $r = 0$ and $r = \pm i$. The solution is

$x(t) = C_1 + C_2 \cos t + C_3 \sin t$.

(c) The characteristic equation $r^4 + 4r^2 + 4 = 0$ can be written $(r^2+2)^2 = 0$ and has repeated roots $r = \pm\sqrt{2}$. The solution is

$x(t) = C_1 \cos \sqrt{2}t + C_2 \sin \sqrt{2}t + C_3 t \cos \sqrt{2}t + C_4 t \sin \sqrt{2}t$.

(e) The characteristic equation $r^4 + 2r^2 - 8 = 0$ can be written $(r^2+1)^2 - 9 = 0$ and so $r^2 = -1 \pm 3$ and has roots $r = \pm\sqrt{2}$ and $r = \pm 2i$. The solution is

$x(t) = C_1 \cos 2t + C_2 \sin 2t + C_3 e^{\sqrt{2}t} + C_4 e^{-\sqrt{2}t}$.

(g) The characteristic equation $r^3 - 2r^2 - r + 2 = 0$ can be written $(r-2)(r-1)(r+1) = 0$ and has roots $r = 1$, $r = 2$, and $r = -1$. The solution is

$x(t) = C_1 e^t + C_2 e^{2t} + C_3 e^{-t}$.

(i) The characteristic equation $r^3 + r^2 + 3r - 5 = 0$ can be written $(r-1)(r^2+2r+5) = 0$ and has roots $r = 1$, $r = -1 \pm 2i$. The solution is

$x(t) = C_1 e^t + C_2 e^{-t} \cos 2t + C_3 e^{-t} \sin 2t$.

3. **(a)** The characteristic equation $r^4 + 3r^3 + 2r^2 = 0$ can be written $r^2(r+1)(r+2) = 0$ and has roots $r = 0$ (repeated), $r = -1$, $r = -2$. The solution is
$x(t) = C_1 + C_2 t + C_3 e^{-t} + C_4 e^{-2t}$.
From this we have
$x'(t) = C_2 - C_3 e^{-t} - 2C_4 e^{-2t}$.
$x''(t) = C_3 e^{-t} + 4C_4 e^{-2t}$.
$x'''(t) = -C_3 e^{-t} - 8C_4 e^{-2t}$.
$x(0) = 0$ gives $C_1 + C_3 + C_4 = 0$, so $C_1 = -C_3 - C_4$.
$x'(0) = 0$ gives $C_2 - C_3 - 2C_4 = 0$, so $C_2 = C_3 + 2C_4$.
$x''(0) = 0$ gives $C_3 + 4C_4 = 0$, so $C_3 = -4C_4$.
$x'''(0) = 8$ gives $-C_3 - 8C_4 = 8$, so $-4C_4 = 8$.
From these we find $C_4 = -2$, $C_3 = 8$, $C_2 = 4$, $C_1 = -6$, so
$x(t) = -6 + 4t + 8e^{-t} - 2e^{-2t}$.

(c) The characteristic equation $r^3 + 6r^2 + 5r - 12 = 0$ can be written $(r-1)(r+3)(r+4) = 0$ and has roots $r = 1$, $r = -3$, $r = -4$. The solution is
$x(t) = C_1 e^t + C_2 e^{-3t} + C_3 e^{-4t}$.
From this we have
$x'(t) = C_1 e^t - 3C_2 e^{-3t} - 4C_3 e^{-4t}$.
$x''(t) = C_1 e^t + 9C_2 e^{-3t} + 16C_3 e^{-4t}$.
$x(0) = 0$ gives $C_1 + C_2 + C_3 = 0$.
$x'(0) = 4$ gives $C_1 - 3C_2 - 4C_3 = 4$.
$x''(0) = -8$ gives $C_1 + 9C_2 + 16C_3 = -8$.
From these we find $C_3 = 0$, $C_2 = -1$, $C_1 = 1$, so
$x(t) = e^t - e^{-3t}$.

5. In each case we compute $W[f_1, f_2, f_3]$. If $W[f_1, f_2, f_3]$ is not identically zero, then the set $\{f_1, f_2, f_3\}$ is linearly independent.

(a) $W = -3$. Linearly independent.

(c) $W = 0$. Linearly dependent, because $t + 3 = 1 \cdot (1+t) + 2 \cdot (1)$.

(e) $W = 2e^{3t}$. Linearly independent.

(g) $W = -6e^{2t}$. Linearly independent.

7. **(a)** $\begin{vmatrix} e^t & e^{2t} & e^{4t} \\ e^t & 2e^{2t} & 4e^{4t} \\ e^t & 4e^{2t} & 16e^{4t} \end{vmatrix} = e^t e^{2t} e^{4t} \begin{vmatrix} 1 & 1 & 1 \\ 1 & 2 & 4 \\ 1 & 4 & 16 \end{vmatrix}$
$= e^{7t}(2-1)(4-1)(4-2) = 6e^{7t}$, which is not identically zero.

(c) $\begin{vmatrix} e^t & e^{2t} & e^{3t} & \cdots & e^{nt} \\ e^t & 2e^{2t} & 3e^{3t} & \cdots & ne^{nt} \\ e^t & 2^2 e^{2t} & 3^2 e^{3t} & \cdots & n^2 e^{nt} \\ \vdots & \vdots & \vdots & \ddots & \vdots \\ e^t & 2^{n-1}e^{2t} & 3^{n-1}e^{3t} & \cdots & n^{n-1}e^{nt} \end{vmatrix}$

$= e^t e^{2t} e^{3t} \cdots e^{nt} \begin{vmatrix} 1 & 1 & 1 & \cdots & 1 \\ 1 & 2 & 3 & \cdots & n \\ 1 & 2^2 & 3^2 & \cdots & n^2 \\ \vdots & \vdots & \vdots & \ddots & \vdots \\ 1 & 2^{n-1} & 3^{n-1} & \cdots & n^{n-1} \end{vmatrix}$

$= e^t e^{2t} e^{3t} \cdots e^{nt} \times$
$\times (2-1)(3-1)(3-2)(4-1)(4-2)(4-3) \cdots (n-1)(n-2) \cdots [n-(n-1)]$,
which is not identically zero.

9. **(a)**

$$\frac{d}{dt}W[f_1,f_2,f_3]=\begin{vmatrix} f_1' & f_2' & f_3' \\ f_1' & f_2' & f_3' \\ f_1'' & f_2'' & f_3'' \end{vmatrix}+\begin{vmatrix} f_1 & f_2 & f_3 \\ f_1'' & f_2'' & f_3'' \\ f_1'' & f_2'' & f_3'' \end{vmatrix}+\begin{vmatrix} f_1 & f_2 & f_3 \\ f_1' & f_2' & f_3' \\ f_1''' & f_2''' & f_3''' \end{vmatrix}.$$

The first two determinants on the right-hand side are each zero, because they each have identical rows.

(b) Now, for $i=1,2$, and 3, $a_3 f_i''' + a_2 f_i'' + a_1 f_i' + a_0 f_i = 0$, so

$$f_i''' = -\frac{a_2}{a_3}f_i'' - \frac{a_1}{a_3}f_i' - \frac{a_0}{a_3}f_i,$$

giving

$$\frac{d}{dt}W[f_1,f_2,f_3]=\begin{vmatrix} f_1 & f_2 & f_3 \\ f_1' & f_2' & f_3' \\ -\frac{a_2}{a_3}f_1''-\frac{a_1}{a_3}f_1'-\frac{a_0}{a_3}f_1 & -\frac{a_2}{a_3}f_2''-\frac{a_1}{a_3}f_2'-\frac{a_0}{a_3}f_2 & -\frac{a_2}{a_3}f_3''-\frac{a_1}{a_3}f_3'-\frac{a_0}{a_3}f_3 \end{vmatrix}$$

Multiplying the first row by $\frac{a_0}{a_3}$ and the second by $\frac{a_1}{a_3}$ and adding the result to the third gives

$$\frac{d}{dt}W[f_1,f_2,f_3]=\begin{vmatrix} f_1 & f_2 & f_3 \\ f_1' & f_2' & f_3' \\ -\frac{a_2}{a_3}f_1'' & -\frac{a_2}{a_3}f_2'' & -\frac{a_2}{a_3}f_3'' \end{vmatrix},$$

or

$$\frac{d}{dt}W[f_1,f_2,f_3]=-\frac{a_2}{a_3}W[f_1,f_2,f_3].$$

(c) The equation $\frac{dy}{dt}=-a(t)\,y$ has solution $y=C\exp\left[-\int a(t)\,dt\right]$, so

$$W[f_1,f_2,f_3]=C\exp\left[-\int\frac{a_2(t)}{a_3(t)}\,dt\right].$$

(d) $W[f_1(t),f_2(t),f_3(t)]=C\exp\left[-\int_{t_0}^{t}\frac{a_2(u)}{a_3(u)}\,du\right]$, so

$$W[f_1(t_0),f_2(t_0),f_3(t_0)]=C.$$

(e) Because

$$W[f_1(t),f_2(t),f_3(t)]=W[f_1(t_0),f_2(t_0),f_3(t_0)]\exp\left[-\int_{t_0}^{t}\frac{a_2(u)}{a_3(u)}\,du\right]$$

then $W[f_1(t_0),f_2(t_0),f_3(t_0)]\neq 0$ is equivalent to $W[f_1(t),f_2(t),f_3(t)]\neq 0$.

11. Differentiating $mx''=k(y-x)$ twice with respect to t gives $mx''''=k(y''-x'')$. Solving $My''=-k(y-x)$ for y'' and substituting in the previous equation gives $mMx''''=k[-k(y-x)-Mx'']$. Substituting for $k(y-x)$ from $mx''=k(y-x)$ gives $mMx''''=k(-mx''-Mx'')$ or, $mMx''''+k(m+M)x''=0$. This fourth order equation has characteristic equation $mMr^4+k(m+M)r^2=0$ with repeated root $r=0$ and complex roots $r=\pm i\omega$, where $\omega^2=k(m+M)/(mM)$. Thus,

$x(t)=At+B+C_1\cos\omega t+C_2\sin\omega t.$

We can find $y(t)$ from $y=x+(m/k)\,x''$, namely,

$y(t)=At+B+C_1\cos\omega t+C_2\sin\omega t+(m/k)\left[-\omega^2C_1\cos\omega t-\omega^2C_2\sin\omega t\right]$

$=At+B+\left(1-m\omega^2/k\right)C_1\cos\omega t+\left(1-m\omega^2/k\right)C_2\sin\omega t$

$=At+B+(1-(m+M)/M)\,C_1\cos\omega t+(1-(m+M)/M)\,C_2\sin\omega t$

$=At+B-(m/M)\,C_1\cos\omega t-(m/M)\,C_2\sin\omega t.$

8.8 Solving Higher Order Nonhomogeneous Differential Equations

1. We list the solution of the associated homogeneous equation, the trial particular solution, and then the explicit solution.

(a) $x_h(t) = C_1 + C_2 \cos t + C_3 \sin t,$
$x_p(t) = A \sin 2t + B \cos 2t,$
$x(t) = \frac{1}{6} \cos 2t + C_1 + C_2 \cos t + C_3 \sin t.$

(c) $x_h(t) = C_1 + C_2 \cos t + C_3 \sin t,$
$x_p(t) = t(A \sin t + B \cos t),$
$x(t) = -\frac{1}{2} t \cos t + C_1 + C_2 \cos t + C_3 \sin t.$

(e) $x_h(t) = C_1 \cos \sqrt{2}t + C_2 \sin \sqrt{2}t + C_3 t \cos \sqrt{2}t + C_4 t \sin \sqrt{2}t,$
$x_p(t) = Ae^{-2t},$
$x(t) = \frac{1}{9} e^{-2t} + C_1 \cos \sqrt{2}t + C_2 \sin \sqrt{2}t + C_3 t \cos \sqrt{2}t + C_4 t \sin \sqrt{2}t.$

(g) $x_h(t) = C_1 + C_2 t + C_3 e^{-t} + C_4 e^{-2t},$
$x_p(t) = Ate^{-t} + Be^{t},$
$x(t) = e^{-t} t + \frac{1}{6} e^{t} + C_1 + C_2 t + C_3 e^{-t} + C_4 e^{-2t}.$

(i) $x_h(t) = C_1 \cos \sqrt{3}t + C_2 \sin \sqrt{3}t + C_3 t \cos \sqrt{3}t + C_4 t \sin \sqrt{3}t,$
$x_p(t) = A \sin t + B \cos t + Ce^{-3t},$
$x(t) = \frac{1}{144} e^{-3t} + \frac{1}{4} \cos t + C_1 \cos \sqrt{3}t + C_2 \sin \sqrt{3}t + C_3 t \cos \sqrt{3}t + C_4 t \sin \sqrt{3}t.$

3. Differentiating $mx'' = mg + \frac{k}{l}(y - x - l)$ twice gives
$mx'''' = \frac{k}{l}(y'' - x'')$ which from $My'' = Mg - \frac{k}{l}(y - x - l)$ can be written
$mMx'''' = \frac{k}{l}\left[Mg - \frac{k}{l}(y - x - l) - Mx''\right].$
Using $\frac{k}{l}(y - x - l) = mx'' - mg$ we have
$mMx'''' = \frac{k}{l}\left[(M + m) g - (M + m) x''\right]$, or
$mMx'''' + \frac{k}{l}(M + m) x'' = \frac{k}{l}\left[(M + m) g\right].$
The homogeneous equation is $mMx'''' + \frac{k}{l}(M + m) x'' = 0$, with characteristic equation $mMr^4 + \frac{k}{l}(M + m) r^2 = 0$ with repeated root $r = 0$ and complex roots $r = \pm i\omega$, where $\omega^2 = k(m + M)/(lmM)$. Thus,
$x_h(t) = C_3 t + C_4 + C_1 \cos \omega t + C_2 \sin \omega t.$
The forcing term $\frac{k}{l}\left[(M + m) g\right]$ leads to a trial solution of $x_p(t) = At^2$, in which case
$mMx_p'''' + \frac{k}{l}(M + m) x_p'' = \frac{k}{l}\left[(M + m) g\right]$ gives
$\frac{k}{l}(M + m) 2A = \frac{k}{l}\left[(M + m) g\right]$, so $A = g/2$, and
$x(t) = gt^2/2 + C_3 t + C_4 + C_1 \cos \omega t + C_2 \sin \omega t.$
We can find $y(t)$ from $mx'' = mg + \frac{k}{l}(y - x - l)$ namely,
$y = x + (lm/k) x'' + l - lmg/k$, so
$y(t) = gt^2/2 + C_3 t + C_4 + C_1 \cos \omega t + C_2 \sin \omega t +$
$\quad + (lm/k)\left[g - \omega^2 C_1 \cos \omega t - \omega^2 C_2 \sin \omega t.\right] + l - lmg/k$
$= gt^2/2 + C_3 t + C_4 + \left(1 - lm\omega^2/k\right) C_1 \cos \omega t + \left(1 - lm\omega^2/k\right) C_2 \sin \omega t + l$
$= gt^2/2 + C_3 t + C_4 - (m/M) C_1 \cos \omega t - (m/M) C_2 \sin \omega t + l.$

9. LINEAR AUTONOMOUS SYSTEMS

9.1 Solving Linear Autonomous Systems

1. **(a)** Characteristic equation

$$\begin{vmatrix} 2-r & 5 \\ 1 & 6-r \end{vmatrix} = (2-r)(6-r) - 5 = r^2 - 8r + 7 = (r-7)(r-1) = 0.$$

Roots 1 and 7, so

$x(t) = C_1e^t + C_2e^{7t}$.

To find $y(t)$ we use

$5y(t) = x' - 2x$

$= C_1e^t + 7C_2e^{7t} - 2\left(C_1e^t + C_2e^{7t}\right)$

$= -C_1e^t + 5C_2e^{7t}$, so

$y(t) = -\frac{1}{5}C_1e^t + C_2e^{7t}$.

(c) Characteristic equation

$$\begin{vmatrix} 5-r & 6 \\ 1 & 4-r \end{vmatrix} = (5-r)(4-r) - 6 = r^2 - 9r + 14 = (r-7)(r-2) = 0.$$

Roots 2 and 7, so

$x(t) = C_1e^{2t} + C_2e^{7t}$.

To find $y(t)$ we use

$6y(t) = x' - 5x$

$= 2C_1e^{2t} + 7C_2e^{7t} - 5\left(C_1e^{2t} + C_2e^{7t}\right)$

$= -3C_1e^t + 2C_2e^{7t}$, so

$y(t) = -\frac{1}{2}C_1e^{2t} + \frac{1}{3}C_2e^{7t}$.

(e) Characteristic equation

$$\begin{vmatrix} 2-r & 9 \\ -1 & -4-r \end{vmatrix} = (2-r)(-4-r) + 9 = r^2 + 2r + 1 = (r+1)^2 = 0.$$

Repeated root of -1, so

$x(t) = C_1e^{-t} + C_2te^{-t}$.

To find $y(t)$ we use

$9y(t) = x' - 2x$

$= -C_1e^{-t} + C_2e^{-t} - C_2te^{-t} - 2\left(C_1e^{-t} + C_2te^{-t}\right)$

$= (-3C_1 + C_2)e^{-t} - 3C_2te^{-t}$, so

$y(t) = \frac{1}{9}(-3C_1 + C_2)e^{-t} - \frac{1}{3}C_2te^{-t}$.

(g) Characteristic equation

$$\begin{vmatrix} 4-r & -2 \\ 5 & -2-r \end{vmatrix} = (4-r)(-2-r) + 10 = r^2 - 2r + 2 = (r-1)^2 + 1 = 0.$$

Roots of $1 \pm i$, so

$x(t) = C_1e^t \sin t + C_2e^t \cos t$.

To find $y(t)$ we use

$-2y(t) = x' - 4x$

$= C_1e^t \sin t + C_1e^t \cos t + C_2e^t \cos t - C_2e^t \sin t - 4\left(C_1e^t \sin t + C_2e^t \cos t\right)$

$= (-3C_1 - C_2)e^t \sin t + (C_1 - 3C_2)e^t \cos t$, so

$y(t) = \frac{1}{2}(3C_1 + C_2)e^t \sin t - \frac{1}{2}(C_1 - 3C_2)e^t \cos t$.

(i) Characteristic equation

$$\begin{vmatrix} 3-r & 5 \\ -5 & 3-r \end{vmatrix} = (3-r)^2 + 25 = 0.$$

Roots of $3 \pm 5i$, so

$x(t) = C_1 e^{3t} \sin 5t + C_2 e^{3t} \cos 5t.$

To find $y(t)$ we use

$5y(t) = x' - 3x$

$= 5C_1 e^{3t} \cos 5t + 3C_1 e^{3t} \sin 5t - 5C_2 e^{3t} \sin 5t + 3C_2 e^{3t} \cos 5t - 3\left(C_1 e^{3t} \sin 5t + C_2 e^{3t} \cos 5t\right)$

$= 5C_1 e^{3t} \cos 5t - 5C_2 e^{3t} \sin 5t$, so

$y(t) = -C_2 e^{3t} \sin 5t + C_1 e^{3t} \cos 5t.$

(k) Characteristic equation

$$\begin{vmatrix} 1-r & 1 \\ -1 & 3-r \end{vmatrix} = (1-r)(3-r) + 1 = r^2 - 4r + 4 = (r-2)^2 = 0.$$

Repeated root of 2, so

$x(t) = (C_1 + C_2 t) e^{2t}.$

To find $y(t)$ we use

$y(t) = x' - x$

$= (2C_1 + 2C_2 t + C_2) e^{2t} - (C_1 + C_2 t) e^{2t}$

$= (C_1 + C_2 + C_2 t) e^{2t}.$

3. **(a)** $2x + 5y - 7 = 0$ and $x + 6y + 14 = 0$ have the solution $x = 16$, $y = -5$, so we introduce the new variables u and v, where $u = x - 16$ and $v = y + 5$. The system of differential equations $x' = 2x + 5y - 7$, $y' = x + 6y + 14$ becomes

$u' = 2(u+16) + 5(v-5) - 7 = 2u + 5v,$

$v' = (u+16) + 6(v-5) + 14 = u + 6v.$

Characteristic equation

$$\begin{vmatrix} 2-r & 5 \\ 1 & 6-r \end{vmatrix} = (2-r)(6-r) - 5 = r^2 - 8r + 7 = (r-1)(r-7) = 0.$$

Roots 1 and 7, so

$u(t) = C_1 e^t + C_2 e^{7t}.$

To find $v(t)$ we use

$5v(t) = u' - 2u$

$= C_1 e^t + 7C_2 e^{7t} - 2\left(C_1 e^t + C_2 e^{7t}\right)$

$= -C_1 e^t + 5C_2 e^{7t}$, so

$v(t) = -\frac{1}{5} C_1 e^t + C_2 e^{7t}$. Thus,

$x(t) = 16 + C_1 e^t + C_2 e^{7t}$

$y(t) = -5 - \frac{1}{5} C_1 e^t + C_2 e^{7t}$

(c) $x - 2y + 1 = 0$ and $2x + 5y - 7 = 0$ have the solution $x = 1$, $y = 1$, so we introduce the new variables u and v, where $u = x - 1$ and $v = y - 1$. The system of differential equations $x' = x - 2y + 1$, $y' = 2x + 5y - 7$ becomes

$u' = (u+1) - 2(v+1) + 1 = u - 2v,$

$v' = 2(u+1) + 5(v+1) - 7 = 2u + 5v.$

Characteristic equation

$$\begin{vmatrix} 1-r & -2 \\ 2 & 5-r \end{vmatrix} = (1-r)(5-r) + 4 = r^2 - 6r + 9 = (r-3)^2 = 0.$$

Repeated root 3, so

$u(t) = (C_1 + C_2 t) e^{3t}.$

To find $v(t)$ we use

$-2v(t) = u' - u$
$= (3C_1 + 3C_2t + C_2)\, e^{3t} - (C_1 + C_2t)\, e^{3t}$
$= (2C_1 + C_2)\, e^t + 2C_2te^t$, so
$v(t) = -\frac{1}{2}(2C_1 + C_2)\, e^t - C_2te^t$. Thus,
$x(t) = 1 + C_1e^{3t} + C_2te^{3t}$
$y(t) = 1 - \frac{1}{2}(2C_1 + C_2)\, e^{3t} - C_2te^{3t}$.

(e) $5x - y + 6 = 0$ and $6x + y + 5 = 0$ have the solution $x = -1$, $y = 1$, so we introduce the new variables u and v, where $u = x + 1$ and $v = y - 1$. The system of differential equations $x' = 5x - y + 6$, $y' = 6x + y + 5$ becomes
$u' = 5(u - 1) - (v + 1) + 6 = 5u - v,$
$v' = 6(u - 1) + (v + 1) + 5 = 6u + v.$
Characteristic equation
$$\begin{vmatrix} 5 - r & -1 \\ 6 & 1 - r \end{vmatrix} = (5 - r)(1 - r) + 6 = r^2 - 6r + 11 = (r - 3)^2 + 2 = 0.$$
Complex roots $3 \pm \sqrt{2}i$, so
$u(t) = e^{3t}\left(C_1 \cos\sqrt{2}t + C_2 \sin\sqrt{2}t\right).$
To find $v(t)$ we use
$-v(t) = u' - 5u$
$= e^{3t}\left(-\sqrt{2}C_1 \sin\sqrt{2}t + \sqrt{2}C_2 \cos\sqrt{2}t + 3C_1 \cos\sqrt{2}t + 3C_2 \sin\sqrt{2}t\right) +$
$\quad -5e^{3t}\left(C_1 \cos\sqrt{2}t + C_2 \sin\sqrt{2}t\right)$
$= e^{3t}\left(-\sqrt{2}C_1 - 2C_2\right) \sin\sqrt{2}t + e^{3t}\left(\sqrt{2}C_2 - 2C_1\right) \cos\sqrt{2}t$, so
$v(t) = e^{3t}\left(\sqrt{2}C_1 + 2C_2\right) \sin\sqrt{2}t + e^{3t}\left(-\sqrt{2}C_2 + 2C_1\right) \cos\sqrt{2}t$. Thus,
$x(t) = -1 + e^{3t}\left(C_1 \cos\sqrt{2}t + C_2 \sin\sqrt{2}t\right)$
$y(t) = 1 + e^{3t}\left(\sqrt{2}C_1 + 2C_2\right) \sin\sqrt{2}t + e^{3t}\left(-\sqrt{2}C_2 + 2C_1\right) \cos\sqrt{2}t.$

5. We let $x(t)$ and $y(t)$ be the amounts of salt in containers A and B respectively at time t. Both x and y have units of pounds, and t has units of minutes. The appropriate differential equations governing this system are obtained from conservation considerations. For container A we have $x' = (4)(3) + (y/100)(1) - (x/100)(4) = 12 - \frac{1}{25}x + \frac{1}{100}y$, whereas for container B we have $y' = \frac{x}{100}4 - \frac{y}{100}1 - \frac{y}{100}3 = \frac{1}{25}x - \frac{1}{25}y$.
$12 + \frac{1}{100}y - \frac{1}{25}x = 0$ and $\frac{1}{25}x - \frac{1}{25}y = 0$ have the solution $x = 400$, $y = 400$, so we introduce the new variables u and v, where $u = x - 400$ and $v = y - 400$. The system of differential equations becomes
$u' = 12 - \frac{1}{25}(u + 400) + \frac{1}{100}(v + 400) = -\frac{1}{25}u + \frac{1}{100}v,$
$v' = \frac{1}{25}(u + 400) - \frac{1}{25}(v + 400) = \frac{1}{25}u - \frac{1}{25}v.$
Characteristic equation
$$\begin{vmatrix} -\frac{1}{25} - r & \frac{1}{100} \\ \frac{1}{25} & -\frac{1}{25} - r \end{vmatrix} = (-1/25 - r)(-1/25 - r) - 1/2500 = 0.$$
Roots $-1/50$ and $-3/50$, so
$u(t) = C_1e^{-t/50} + C_2e^{-3t/50}.$
To find $v(t)$ we use
$\frac{1}{100}v = u' + \frac{1}{25}u$
$= -\frac{1}{50}C_1e^{-t/50} - \frac{3}{50}C_2e^{-3t/50} + \frac{1}{25}\left(C_1e^{-t/50} + C_2e^{-3t/50}\right)$
$= \frac{1}{50}C_1e^{-t/50} - \frac{1}{50}C_2e^{-3t/50}$, so
$v(t) = 2C_1e^{-t/50} - 2C_2e^{-3t/50}$. Thus,
$x(t) = 400 + C_1e^{-t/50} + C_2e^{-3t/50}$
$y(t) = 400 + 2C_1e^{-t/50} - 2C_2e^{-3t/50}$

(a) If there is no salt in either container, the initial conditions at $t = 0$ are $x(0) = y(0) = 0$. If we substitute these values into $x(t)$ and $y(t)$ we obtain $0 = 400 + C_1 + C_2$ and $0 = 400 + 2C_1 - 2C_2$, with solution $C_1 = -300$ and $C_2 = -100$. In this case the amount of salt in containers A and B at time t would be
$x(t) = 400 - 300e^{-t/50} - 100e^{-3t/50}$, $y(t) = 400 - 600e^{-t/50} + 200e^{-3t/50}$.

(b) If this process started with no salt in container A but 300 pounds in container B, the initial conditions would be $x(0) = 0$ and $y(0) = 300$. If we substitute these values into $x(t)$ and $y(t)$ we obtain $0 = 400 + C_1 + C_2$ and $300 = 400 + 2C_1 - 2C_2$, with solution $C_1 = -225$ and $C_2 = -175$. In this case the amount of salt in containers A and B at time t would be
$x(t) = 400 - 225e^{-t/50} - 175e^{-3t/50}$, $y(t) = 400 - 450e^{-t/50} + 350e^{-3t/50}$.

9.2 Classification of Solutions via Stability

1. Notice that these equations are the same as the ones in Exercise 1 of Section 9.1. Also, in each case, the origin is the only equilibrium point. We need only know the roots of the characteristic equation and then use the Classification Theorem.

(a) $r_1 = 1$, $r_2 = 7$. $0 < r_1 < r_2$, so the origin is an unstable node.

(c) $r_1 = 2$, $r_2 = 7$. $0 < r_1 < r_2$, so the origin is an unstable node.

(e) $r_1 = r_2 = -1 < 0$, so the origin is a stable node.

(g) $r_{1,2} = 1 \pm i$. $\alpha = 1 > 0$, so the origin is an unstable focus.

(i) $r_{1,2} = 3 \pm 5i$. $\alpha = 3 > 0$, so the origin is an unstable focus.

(k) $r_1 = r_2 = 2 > 0$, so the origin is an unstable node.

3. (a) $x' = -x, \Longrightarrow x = C_1e^{-t}$. $y' = -2y, \Longrightarrow y = C_2e^{-2t}$. Thus, the origin is a stable node.

(b) In the phase plane, $dy/dx = -2y/(-x)$, or $(1/y)\,dy/dx = 2/x$. This gives $\ln|y| = 2\ln|x| + C$, or $y = \pm e^C x^2$, and we have nodes. Checking the signs of x' and y' shows the origin is a stable node.

9.3 When Do Straight-Line Orbits Exist?

1. Notice that these equations are the same as the ones in Exercise 1 of Sections 9.1 and 9.2.

(a) The differential equation in the phase plane is

$$\frac{dy}{dx} = \frac{x + 6y}{2x + 5y},$$

which has a straight-line orbit, $y = mx$, if

$$m = \frac{1 + 6m}{2 + 5m}.$$

This gives the quadratic equation $5m^2 - 4m - 1 = 0$, with solutions $m = 1, -\frac{1}{5}$. Two straight-line orbits $y = x$, $y = -\frac{1}{5}x$.

(c) The differential equation in the phase plane is

$$\frac{dy}{dx} = \frac{x+4y}{5x+6y},$$

which has a straight-line orbit, $y = mx$, if

$$m = \frac{1+4m}{5+6m}.$$

This gives the quadratic equation $6m^2 + m - 1 = 0$, with solutions $m = -\frac{1}{2}, \frac{1}{3}$. Two straight-line orbits $y = \frac{1}{3}x$, $y = -\frac{1}{2}x$.

(e) The differential equation in the phase plane is

$$\frac{dy}{dx} = \frac{-x-4y}{2x+9y},$$

which has a straight-line orbit, $y = mx$, if

$$m = \frac{-1-4m}{2+9m}.$$

This gives the quadratic equation $9m^2 + 6m + 1 = 0$, with solution $m = -\frac{1}{3}$. One straight-line orbit $y = -\frac{1}{3}x$.

(g) The differential equation in the phase plane is

$$\frac{dy}{dx} = \frac{5x-2y}{4x-2y},$$

which has a straight-line orbit, $y = mx$, if

$$m = \frac{5-2m}{4-2m}.$$

This gives the quadratic equation $2m^2 - 6m + 5 = 0$, with no real solutions. No straight-line orbits.

(i) The differential equation in the phase plane is

$$\frac{dy}{dx} = \frac{-5x+3y}{3x+5y},$$

which has a straight-line orbit, $y = mx$, if

$$m = \frac{-5+3m}{3+5m}.$$

This gives the quadratic equation $5m^2+5 = 0$, with no real solutions. No straight-line orbits.

(k) The differential equation in the phase plane is

$$\frac{dy}{dx} = \frac{-x+3y}{x+y},$$

which has a straight-line orbit, $y = mx$, if

$$m = \frac{-1+3m}{1+m}.$$

This gives the quadratic equation $m^2 - 2m + 1 = 0$, with solution $m = 1$. One straight-line orbit $y = x$.

3. (a) Figure 9.18 of the text has two straight-line orbits ($y = \pm x$).

(b) Figure 9.20 of the text has no straight-line orbits.

(c) Figure 9.19 of the text has one straight-line orbit ($y = x$).

9.4 Qualitative Behavior Using Nullclines

1. (a) The x-nullcline (vertical arrows) is $y = -2x/5$, and the y-nullcline (horizontal arrows) is $y = -x/6$.

    The vertical arrows point up if $y' = x + 6y$ is positive when $y = -2x/5$, that is $y' = x + 6(-2x/5) = -7x/5$. So the vertical arrows point up when $x < 0$ and down when $x > 0$.

    The horizontal arrows point right if $x' = 2x + 5y$ is positive when $y = -x/6$, that is $x' = 2x + 5(-x/6) = 7x/6$. So the horizontal arrows point right when $x > 0$ and left when $x < 0$.

    These nullclines indicate that all orbits are leaving the origin without spiralling, so $(0,0)$ is an unstable node. See Figure 9-1.

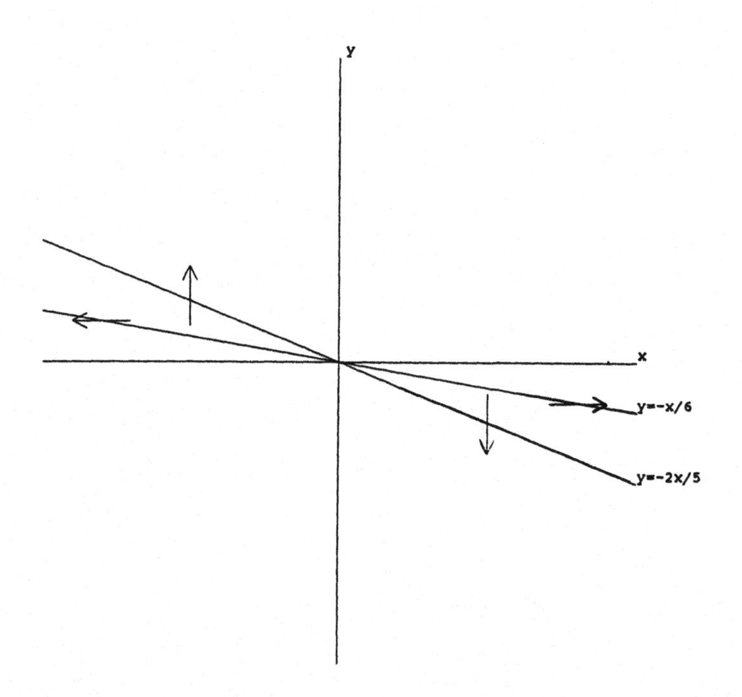

Figure 9-1 Exercise 1(a).

(c) The x-nullcline (vertical arrows) is $y = -5x/6$, and the y-nullcline (horizontal arrows) is $y = -x/4$.

The vertical arrows point up if $y' = x + 4y$ is positive when $y = -5x/6$, that is $y' = x + 4(-5x/6) = -7x/2$. So the vertical arrows point up when $x < 0$ and down when $x > 0$.

The horizontal arrows point right if $x' = 5x + 6y$ is positive when $y = -x/4$, that is $x' = 5x + 6(-x/4) = 7x/2$. So the horizontal arrows point right when $x > 0$ and left when $x < 0$.

These nullclines indicate that all orbits are leaving the origin without spiralling, so $(0,0)$ is an unstable node. See Figure 9-2.

(e) The x-nullcline (vertical arrows) is $y = -2x/9$, and the y-nullcline (horizontal arrows) is $y = -x/4$.

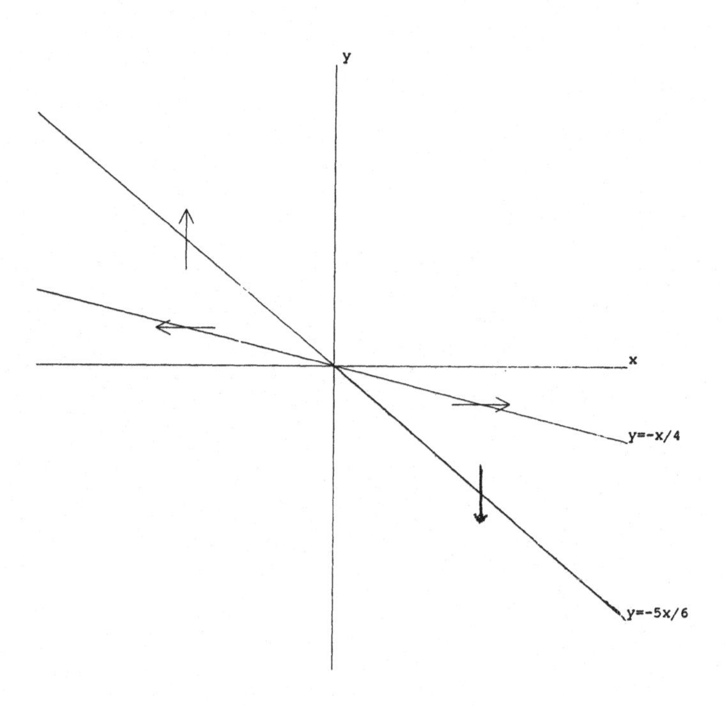

Figure 9-2 Exercise 1(c).

The vertical arrows point up if $y' = -x - 4y$ is positive when $y = -2x/9$, that is $y' = -x - 4(-2x/9) = -x/9$. So the vertical arrows point up when $x < 0$ and down when $x > 0$.

The horizontal arrows point right if $x' = 2x + 9y$ is positive when $y = -x/4$, that is $x' = 2x + 9(-x/4) = -x/4$. So the horizontal arrows point right when $x < 0$ and left when $x > 0$.

The arrows on the nullclines all point in a clockwise direction, so the nullcline analysis is inconclusive. See Figure 9-3.

(g) The x-nullcline (vertical arrows) is $y = 2x$, and the y-nullcline (horizontal arrows) is $y = 5x/2$.

The vertical arrows point up if $y' = 5x - 2y$ is positive when $y = 2x$, that is $y' = 5x - 2(2x) = x$. So the vertical arrows point up when $x > 0$ and down when $x < 0$.

The horizontal arrows point right if $x' = 4x - 2y$ is positive when $y = 5x/2$, that is $x' = 4x - 2(5x/2) = -x$. So the horizontal arrows point right when $x < 0$ and left when $x > 0$.

The arrows on the nullclines all point in a clockwise direction, so the nullcline analysis is inconclusive. See Figure 9-4.

(i) The x-nullcline (vertical arrows) is $y = -3x/5$, and the y-nullcline (horizontal arrows) is $y = 5x/3$.

The vertical arrows point up if $y' = -5x + 3y$ is positive when $y = -3x/5$, that is $y' = -5x + 3(-3x/5) = -34x/5$. So the vertical arrows point up when $x < 0$ and down when $x > 0$.

The horizontal arrows point right if $x' = 3x + 5y$ is positive when $y = 5x/3$, that is $x' = 3x + 5(5x/3) = 34x/3$. So the horizontal arrows point right when $x > 0$ and left when $x < 0$.

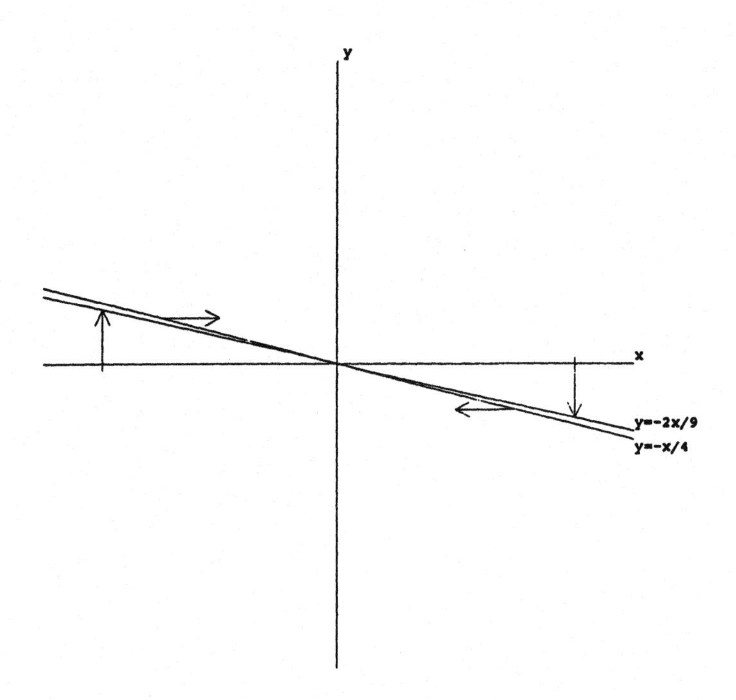

Figure 9-3 Exercise 1(e).

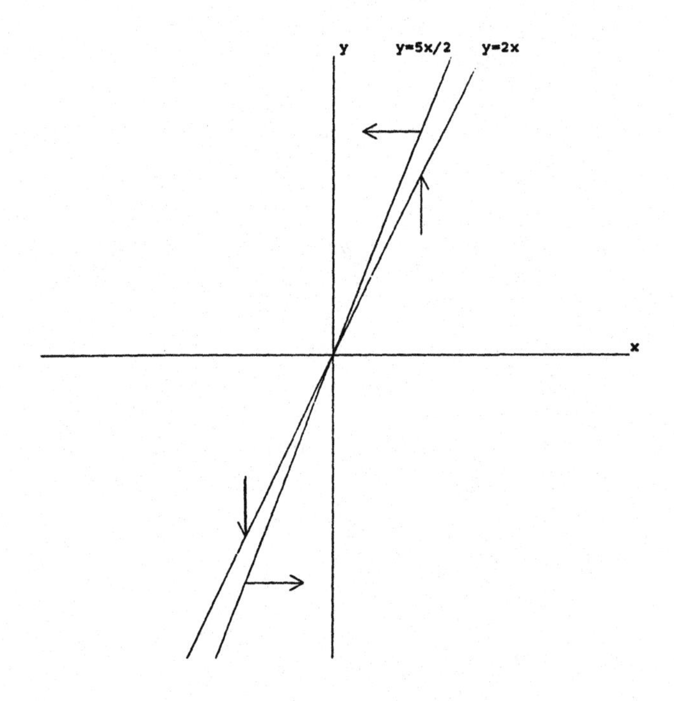

Figure 9-4 Exercise 1(g).

The arrows on the nullclines all point in a clockwise direction, so the nullcline analysis is inconclusive. See Figure 9-5.

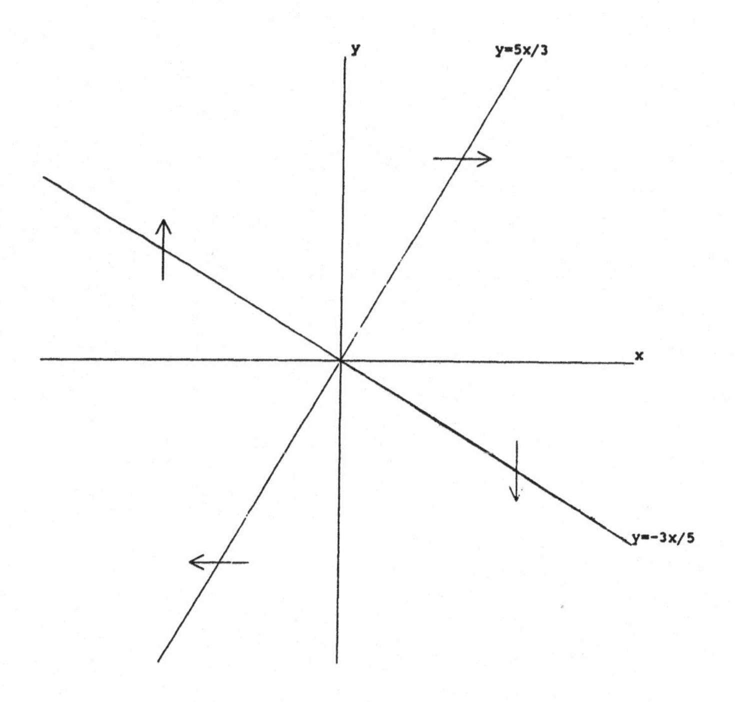

Figure 9-5 Exercise 1(i).

(k) The x-nullcline (vertical arrows) is $y = -x$, and the y-nullcline (horizontal arrows) is $y = x/3$.

The vertical arrows point up if $y' = -x + 3y$ is positive when $y = -x$, that is $y' = -x + 3(-x) = -4x$. So the vertical arrows point up when $x < 0$ and down when $x > 0$.

The horizontal arrows point right if $x' = x + y$ is positive when $y = x/3$, that is $x' = x + x/3 = 4x/3$. So the horizontal arrows point right when $x > 0$ and left when $x < 0$.

The arrows on the nullclines all point in a clockwise direction, so the nullcline analysis is inconclusive. See Figure 9-6.

9.5 Matrix Formulation of Solutions

1. **(a)** $\mathbf{V} = \begin{bmatrix} A \\ B \end{bmatrix} f(t) = \begin{bmatrix} Af(t) \\ Bf(t) \end{bmatrix}$, so

$$\mathbf{V}' = \begin{bmatrix} Af(t) \\ Bf(t) \end{bmatrix}' = \begin{bmatrix} Af'(t) \\ Bf'(t) \end{bmatrix} = \begin{bmatrix} A \\ B \end{bmatrix} f'(t).$$

(b) $\mathbf{V_1} + \mathbf{V_2} = \begin{bmatrix} x_1(t) \\ y_1(t) \end{bmatrix} + \begin{bmatrix} x_2(t) \\ y_2(t) \end{bmatrix} = \begin{bmatrix} x_1(t) + x_2(t) \\ y_1(t) + y_2(t) \end{bmatrix}$, so

$$(\mathbf{V_1} + \mathbf{V_2})' = \begin{bmatrix} x_1(t) + x_2(t) \\ y_1(t) + y_2(t) \end{bmatrix}'$$

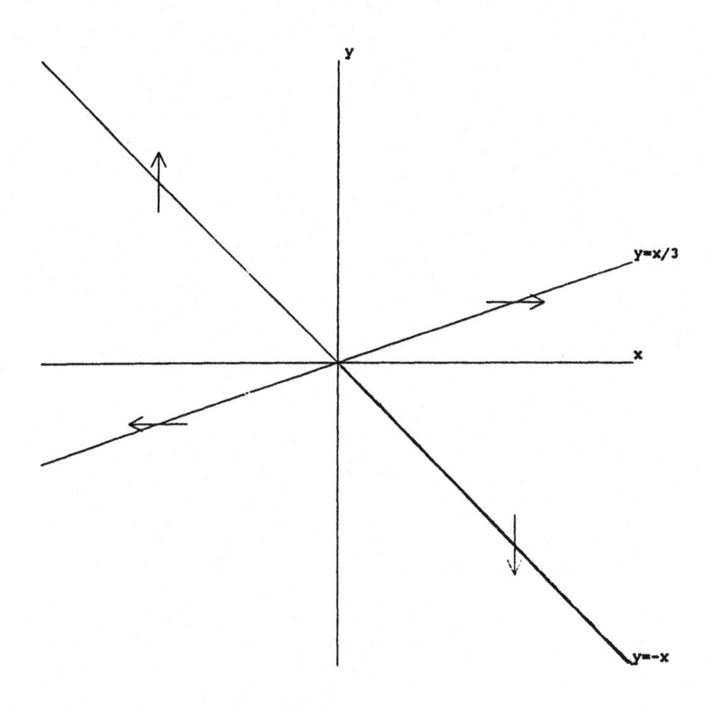

Figure 9-6 Exercise 1(k).

$$= \begin{bmatrix} x_1'(t) + x_2'(t) \\ y_1'(t) + y_2'(t) \end{bmatrix} = \begin{bmatrix} x_1'(t) \\ y_1'(t) \end{bmatrix} + \begin{bmatrix} x_2'(t) \\ y_2'(t) \end{bmatrix} = \mathbf{V}_1' + \mathbf{V}_2'.$$

$$(z(t)\mathbf{V})' = \begin{bmatrix} z(t)x_1(t) \\ z(t)y_1(t) \end{bmatrix}'$$

$$= \begin{bmatrix} z'(t)x_1(t) + z(t)x_1'(t) \\ z'(t)y_1(t) + z(t)y_1'(t) \end{bmatrix} = \begin{bmatrix} z'(t)x_1(t) \\ z'(t)y_1(t) \end{bmatrix} + \begin{bmatrix} z(t)x_1'(t) \\ z(t)y_1'(t) \end{bmatrix} = z'(t)\mathbf{V}+z(t)\mathbf{V}'.$$

3. **(a)** $\mathbf{MZ}_2 = r\mathbf{Z}_2$.

(b) $\mathbf{MX} = \mathbf{X}'(t)$ gives

$$\mathbf{M}\left(\begin{bmatrix} A_1 \\ B_1 \end{bmatrix} e^{rt} + \mathbf{Z}_2 te^{rt}\right) = \begin{bmatrix} A_1 \\ B_1 \end{bmatrix} re^{rt} + \mathbf{Z}_2 e^{rt} + r\mathbf{Z}_2 te^{rt} \text{ so that}$$

$$\mathbf{M}\begin{bmatrix} A_1 \\ B_1 \end{bmatrix} = \begin{bmatrix} A_1 \\ B_1 \end{bmatrix} r + \mathbf{Z}_2 \text{ and}$$

$\mathbf{MZ}_2 = r\mathbf{Z}_2$, that is

$$(\mathbf{M} - \mathbf{rI})\begin{bmatrix} A_1 \\ B_1 \end{bmatrix} = \mathbf{Z}_2 \text{ and the equation from part (a).}$$

5. **(a)** The characteristic equation is $\begin{vmatrix} 2-r & -1 \\ -6 & 1-r \end{vmatrix} = 0.$

Thus, $(2-r)(1-r)-6 = 0$, or $r^2 - 3r - 4 = (r-4)(r+1) = 0$. The eigenvalues are 4 and -1.

For $r = 4$,

$$\begin{bmatrix} -2 & -1 \\ -6 & -3 \end{bmatrix}\begin{bmatrix} A \\ B \end{bmatrix} = \begin{bmatrix} 0 \\ 0 \end{bmatrix} \Longrightarrow -2A = B,$$

so solutions are multiples of $\begin{bmatrix} 1 \\ -2 \end{bmatrix} e^{4t}$.

For $r = -1$,

$$\begin{bmatrix} 3 & -1 \\ -6 & 2 \end{bmatrix}\begin{bmatrix} A \\ B \end{bmatrix} = \begin{bmatrix} 0 \\ 0 \end{bmatrix} \Longrightarrow 3A = B,$$

so solutions are multiples of $\begin{bmatrix} 1 \\ 3 \end{bmatrix} e^{-t}$.

A fundamental matrix is $\mathbf{U}(t) = \begin{bmatrix} e^{4t} & e^{-t} \\ -2e^{4t} & 3e^{-t} \end{bmatrix}$, and $\mathbf{X}(t) = \mathbf{U}(t)\begin{bmatrix} C_1 \\ C_2 \end{bmatrix}$.

The differential equation in the phase plane is

$$\frac{dy}{dx} = \frac{-6x + y}{2x - y},$$

which has a straight-line orbit, $y = mx$, if

$$m = \frac{-6 + m}{2 - m}.$$

This gives the quadratic equation $m^2 - m - 6 = 0$, with solutions $m = 3, -2$, which agree with the eigenvectors.

(c) The characteristic equation is $\begin{vmatrix} 5 - r & -3 \\ -1 & -1 - r \end{vmatrix} = 0$.

Thus, $(5 - r)(-1 - r) - 3 = 0$, or $r^2 - 4r - 8 = 0$ with eigenvalues $2 \pm 2\sqrt{3}$.

For $r = 2 + 2\sqrt{3}$,

$$\begin{bmatrix} 3 - 2\sqrt{3} & -3 \\ -1 & -3 - 2\sqrt{3} \end{bmatrix}\begin{bmatrix} A \\ B \end{bmatrix} = \begin{bmatrix} 0 \\ 0 \end{bmatrix} \Longrightarrow B = \tfrac{3 - 2\sqrt{3}}{3} A,$$

so solutions are multiples of $\begin{bmatrix} 3 \\ 3 - 2\sqrt{3} \end{bmatrix} e^{(2+2\sqrt{3})t}$.

For $r = 2 - 2\sqrt{3}$,

$$\begin{bmatrix} 3 + 2\sqrt{3} & -3 \\ -1 & -3 + 2\sqrt{3} \end{bmatrix}\begin{bmatrix} A \\ B \end{bmatrix} = \begin{bmatrix} 0 \\ 0 \end{bmatrix} \Longrightarrow B = \tfrac{3 + 2\sqrt{3}}{3} A,$$

so solutions are multiples of $\begin{bmatrix} 3 \\ 3 + 2\sqrt{3} \end{bmatrix} e^{(2-2\sqrt{3})t}$.

A fundamental matrix is $\mathbf{U}(t) = \begin{bmatrix} 3e^{(2+2\sqrt{3})t} & 3e^{(2-2\sqrt{3})t} \\ (3 - 2\sqrt{3})\, e^{(2+2\sqrt{3})t} & (3 + 2\sqrt{3})\, e^{(2-2\sqrt{3})t} \end{bmatrix}$,

and $\mathbf{X}(t) = \mathbf{U}(t)\begin{bmatrix} C_1 \\ C_2 \end{bmatrix}$.

The differential equation in the phase plane is

$$\frac{dy}{dx} = \frac{-x - y}{5x - 3y},$$

which has a straight-line orbit, $y = mx$, if

$$m = \frac{-1 - m}{5 - 3m}.$$

This gives the quadratic equation $3m^2 - 6m - 1 = 0$, with solutions $m = (3 \pm 2\sqrt{3})/3$, which agree with the eigenvectors.

(e) The characteristic equation is $\begin{vmatrix} 2 - r & 9 \\ -1 & -4 - r \end{vmatrix} = 0$.

Thus, $(2 - r)(-4 - r) + 9 = 0$, with repeated eigenvalue -1.

We seek a solution of the form

$$\mathbf{X}(t) = \left(\begin{bmatrix} A_1 \\ B_1 \end{bmatrix} + \begin{bmatrix} A_2 \\ B_2 \end{bmatrix} t\right) e^{-t}.$$

Substituting this expression into the differential equation gives

$$\begin{bmatrix} -A_1 \\ -B_1 \end{bmatrix} e^{-t} + \begin{bmatrix} (-t+1)A_2 \\ (-t+1)B_2 \end{bmatrix} e^{-t} = \begin{bmatrix} 2A_1 + 9B_1 + t(2A_2 + 9B_2) \\ -A_1 - 4B_1 + t(-A_2 - 4B_2) \end{bmatrix} e^{-t}.$$

Because we want this equation to be valid for all values of t we equate coefficients of e^t and te^t and obtain the system of algebraic equations

$$\begin{aligned} -A_1 + A_2 &= 2A_1 + 9B_1, \\ -B_1 + B_2 &= -A_1 - 4B_1, \\ -A_2 &= 2A_2 + 9B_2, \\ -B_2 &= -A_2 - 4B_2. \end{aligned}$$

This system of equations has the solution

$$\begin{aligned} A_1 &= -3B_1 - B_2, \\ A_2 &= -3B_2, \end{aligned}$$

where B_1 and B_2 may be chosen arbitrarily, so we denote them by C_1 and C_2. This means that the solution may be written in the form

$$\mathbf{X}(t) = \begin{bmatrix} -3C_1 - C_2 \\ C_1 \end{bmatrix} e^{-t} + \begin{bmatrix} -3C_2 \\ C_2 \end{bmatrix} te^{-t},$$

or

$$\mathbf{X}(t) = \begin{bmatrix} -3e^{-t} & (-3t-1)e^{-t} \\ e^{-t} & te^{-t} \end{bmatrix} \begin{bmatrix} C_1 \\ C_2 \end{bmatrix}.$$

Here a fundamental matrix is

$$\mathbf{U} = \begin{bmatrix} -3e^{-t} & (-3t-1)e^{-t} \\ e^{-t} & te^{-t} \end{bmatrix}.$$

The differential equation in the phase plane is

$$\frac{dy}{dx} = \frac{-x - 4y}{2x + 9y},$$

which has a straight-line orbit, $y = mx$, if

$$m = \frac{-1 - 4m}{2 + 9m}.$$

This gives the quadratic equation $9m^2 + 6m + 1 = 0$, with solution $m = -1/3$, which agree with the eigenvectors.

(g) The characteristic equation is $\begin{vmatrix} 2-r & 1 \\ -4 & 2-r \end{vmatrix} = 0.$

Thus, $(2-r)^2 + 4 = 0$, with eigenvalues $2 \pm 2i$.

For $r = 2 + 2i$,

$$\begin{bmatrix} -2i & 1 \\ -4 & -2i \end{bmatrix} \begin{bmatrix} A \\ B \end{bmatrix} = \begin{bmatrix} 0 \\ 0 \end{bmatrix} \Longrightarrow B = 2Ai,$$

so we have an eigenvector $\begin{bmatrix} 1 \\ 2i \end{bmatrix} e^{(2+2i)t}$.

However,

$$\begin{bmatrix} 1 \\ 2i \end{bmatrix} e^{(2+2i)t}$$

$$= e^{2t} \begin{bmatrix} 1 \\ 2i \end{bmatrix} (\cos 2t + i \sin 2t)$$

$$= e^{2t}\begin{bmatrix} \cos 2t + i\sin 2t \\ -2\sin 2t + 2i\cos 2t \end{bmatrix}$$

$$= e^{2t}\begin{bmatrix} \cos 2t \\ -2\sin 2t \end{bmatrix} + ie^{2t}\begin{bmatrix} \sin 2t \\ 2\cos 2t \end{bmatrix},$$

so a fundamental matrix is

$$\mathbf{U}(t) = \begin{bmatrix} e^{2t}\cos 2t & e^{2t}\sin 2t \\ -2e^{2t}\sin 2t & 2e^{2t}\cos 2t \end{bmatrix}, \text{ and } \mathbf{X}(t) = \mathbf{U}(t)\begin{bmatrix} C_1 \\ C_2 \end{bmatrix}.$$

The differential equation in the phase plane is

$$\frac{dy}{dx} = \frac{-4x + 2y}{2x + y},$$

which has a straight-line orbit, $y = mx$, if

$$m = \frac{-4 + 2m}{2 + m}.$$

This gives the quadratic equation $m^2 + 4 = 0$, which has only complex roots $m = \pm 2i$, which agree with the eigenvectors.

(i) The characteristic equation is $\begin{vmatrix} 3-r & 5 \\ -5 & 3-r \end{vmatrix} = (3-r)^2 + 25 = 0$ with roots $3 \pm 5i$.

For $r = 3 + 5i$,

$$\begin{bmatrix} -5i & 5 \\ -5 & -5i \end{bmatrix}\begin{bmatrix} A \\ B \end{bmatrix} = \begin{bmatrix} 0 \\ 0 \end{bmatrix} \Longrightarrow B = Ai.$$

so we have an eigenvector $\begin{bmatrix} 1 \\ i \end{bmatrix} e^{(3+5i)t}$.

However,

$$\begin{bmatrix} 1 \\ i \end{bmatrix} e^{(3+5i)t}$$

$$= e^{3t}\begin{bmatrix} 1 \\ i \end{bmatrix}(\cos 5t + i\sin 5t)$$

$$= e^{3t}\begin{bmatrix} \cos 5t + i\sin 5t \\ i\cos 5t - \sin 5t \end{bmatrix}$$

$$= e^{3t}\begin{bmatrix} \cos 5t \\ -\sin 5t \end{bmatrix} + ie^{3t}\begin{bmatrix} \sin 5t \\ \cos 5t \end{bmatrix},$$

so a fundamental matrix is

$$\mathbf{U}(t) = \begin{bmatrix} e^{3t}\cos 5t & e^{3t}\sin 5t \\ -e^{3t}\sin 5t & e^{3t}\cos 5t \end{bmatrix}, \text{ and } \mathbf{X}(t) = \mathbf{U}(t)\begin{bmatrix} C_1 \\ C_2 \end{bmatrix}.$$

The differential equation in the phase plane is

$$\frac{dy}{dx} = \frac{-5x + 3y,}{3x + 5y}$$

which has a straight-line orbit, $y = mx$, if

$$m = \frac{-5 + 3m}{3 + 5m}.$$

This gives the quadratic equation $m^2 + 1 = 0$, which has only complex roots $m = \pm i$, which agree with the eigenvectors.

7. The solution is given $\mathbf{X}(t) = \begin{bmatrix} e^{-t/50} & e^{-3t/50} \\ 2e^{-t/50} & -2e^{-3t/50} \end{bmatrix}\begin{bmatrix} C_1 \\ C_2 \end{bmatrix}$.

(a) Initially $x(0) = 10$ and $y(0) = 0$, so

$$\begin{bmatrix} 1 & 1 \\ 2 & -2 \end{bmatrix} \begin{bmatrix} C_1 \\ C_2 \end{bmatrix} = \begin{bmatrix} 10 \\ 0 \end{bmatrix}, \text{ which gives } C_1 = C_2 = 5.$$

We will have $x(t) = y(t)$ when

$5e^{-t/50} + 5e^{-3t/50} = 10e^{-t/50} - 10e^{-3t/50}$, that is, when

$15e^{-3t/50} = 5e^{-t/50}$, or

$e^{2t/50} = 3$, so

$t = 25 \ln 3 \approx 27.5$ min.

(b) $\mathbf{X}(25) = \begin{bmatrix} e^{-1/2} & e^{-3/2} \\ 2e^{-1/2} & -2e^{-3/2} \end{bmatrix} \begin{bmatrix} 5 \\ 5 \end{bmatrix} = \begin{bmatrix} 5\left(e^{-1/2} + e^{-3/2}\right) \\ 10\left(e^{-1/2} - e^{-3/2}\right) \end{bmatrix} \approx \begin{bmatrix} 4.1 \\ 3.8 \end{bmatrix}$ lbs.

(c) As $t \to \infty$, $\mathbf{X}(t) \to \begin{bmatrix} 0 \\ 0 \end{bmatrix}$. Both eigenvalues are negative so the solutions will tend to 0 as $t \to \infty$.

9. Mystery direction field A has no straight-line solutions, so corresponds to case (c). Mystery direction field B has two straight-line solutions, so corresponds to case (a). The slopes of these straight lines appear to be 1 and -1, which are then the ratio of eigenvector components. Mystery direction field C has one straight-line solution, so corresponds to case (b). The slope of this straight line appears to be 1, which would then be the ratio of eigenvector components.

9.6 Compartmental Models

1. **(a)** $dy/dx = -v/V$. This will give straight-line orbits with slope $-v/V$. Increasing v/V makes the orbits more vertical.

(b) The orbits are given by $y(t) = -vx(t)/V + C$. When $t = L$ we have $0 = -vx(L)/V + C$ so $C = vx(L)/V$, and $y(t) = -v[x(t) - x(L)]/V$. As the distance t increases, $x' > 0$ and $y' < 0$ if $x - y < 0$, and $x' < 0$ and $y' > 0$ if $x - y > 0$. Thus, if $y > x$, the orbits will move to the right. If $y < x$, the orbits will move to the left.

(c) Let $w_1 = a/v$ and $w_2 = a/V$, so $\mathbf{M} = \begin{bmatrix} -w_1 & w_1 \\ w_2 & -w_2 \end{bmatrix}$.

The characteristic equation is $(-w_1 - r)(-w_2 - r) - w_1 w_2 = 0$, with roots $r = 0$ and $r = -(w_1 + w_2)$. Thus,

$x(t) = C_1 + C_2 e^{-(w_1+w_2)t}$ and $y(t) = C_1 - \frac{w_2}{w_1} C_2 e^{-(w_1+w_2)t}$.

$y(L) = 0$ gives $C_1 = \frac{w_2}{w_1} C_2 e^{-(w_1+w_2)L}$, and $x(0) = x_0$ gives $C_1 + C_2 = x_0$ so

$C_2 = w_1 x_0 / \left[w_1 + w_2 e^{-(w_1+w_2)L}\right]$ and $C_1 = w_2 e^{-(w_1+w_2)L} x_0 / \left[w_1 + w_2 e^{-(w_1+w_2)L}\right]$.

10. NONLINEAR AUTONOMOUS SYSTEMS

10.1 Introduction to Nonlinear Autonomous Equations

1. The orbits in the phase plane satisfy

$$\frac{dy}{dx} = \frac{\frac{dy}{dt}}{\frac{dx}{dt}} = \frac{y-y^2}{-x} = \frac{y(y-1)}{x},$$

a separable differential equation.

The equilibrium solutions are $y(x) = 0$ and $y(x) = 1$.

The nonequilibrium solutions are obtained from

$$\int \frac{1}{y(y-1)}\, dy = \int \frac{1}{x}\, dx.$$

By partial fractions we have

$$\int \frac{1}{y(y-1)}\, dy = \int \left(\frac{1}{y} - \frac{1}{y-1}\right) dy = \ln|y| - \ln|y-1|,$$

so $\ln|y| - \ln|y-1| = \ln|x| + C$, or $\frac{y}{y-1} = cx$. Solving for y gives

$$y = \frac{cx}{cx-1}.$$

3. **(a)** Constant solutions will satisfy $-x = 0$, $y - x^2 = 0$, that is $x = 0$ and $y = 0$.

(b) i. From $x' = -x$ we have $x(t) = K_1 e^{-t}$, which when substituted into $y' = y - x^2$ gives the linear differential equation

$y' - y = -K_1^2 e^{-2t}$.

This has integrating factor $\mu = e^{-t}$, and so

$(e^{-t}y)' = -K_1^2 e^{-3t}$, which gives

$e^{-t}y = \frac{1}{3}K_1^2 e^{-3t} + K_2$, or

$y(t) = \frac{1}{3}K_1^2 e^{-2t} + K_2 e^t$.

ii. From $y' = y - x^2$ we have

$y'' = y' - 2xx' = y' + 2x^2$. Substituting $y' = y - x^2$ in the form $x^2 = -y' + y$ in this last equation gives

$y'' = y' + 2(-y' + y) = -y' + 2y$, or

$y'' + y' - 2y = 0$.

This is a second order linear differential equation with characteristic equation $r^2 + r - 2 = 0$, with roots -2 and 1, so

$y(t) = C_1 e^{-2t} + C_2 e^t$. This is the same answer as in part (i), with $C_1 = \frac{1}{3}K_1^2$ and $C_2 = K_2$.

If we substitute $y(t)$ and its derivative into $x^2 = -y' + y$ we find

$x^2 = -\left(-2C_1 e^{-2t} + C_2 e^t\right) + \left(C_1 e^{-2t} + C_2 e^t\right) = 3C_1 e^{-2t} = K_1^2 e^{-2t}$, so

$x(t) = K_1 e^{-t}$. This agrees with part (i).

(c) As $t \to \infty$, $x(t) = K_1 e^{-t} \to 0$, and $y(t) = \frac{1}{3}K_1^2 e^{-2t} + K_2 e^t \to \infty$ if $K_2 \neq 0$ and $y(t) \to 0$ if $K_2 = 0$.

(d) i. From part (b) we have $x(t) = K_1 e^{-t}$ and $y(t) = \frac{1}{3}K_1^2 e^{-2t} + K_2 e^t$. The first equation gives $e^t = K_1/x$, which when substituted in the second gives
$y = \frac{1}{3}K_1^2 (K_1/x)^{-2} + K_2 K_1/x = \frac{1}{3}x^2 + K_2 K_1/x$, if $K_1 \neq 0$.
If $K_1 = 0$, then $x = 0$ and $y = K_2 e^t$.
ii. The orbits in the phase plane satisfy

$$\frac{dy}{dx} = \frac{\frac{dy}{dt}}{\frac{dx}{dt}} = \frac{y - x^2}{-x},$$

or

$$\frac{dy}{dx} + \frac{1}{x}y = x,$$

a linear differential equation. This has integrating factor
$\mu = e^{\int (1/x)\,dx} = e^{\ln x} = x$ and so
$\frac{d}{dx}(xy) = x^2$, which gives $xy = \frac{1}{3}x^3 + C$, or
$y = \frac{1}{3}x^2 + C/x$.

(e) The x-nullcline is $x = 0$, and the y-nullcline is $y = x^2$. The vertical arrows on the x-nullcline point up for $y > 0$ and down for $y < 0$ (because, when $x = 0$, $y' = y$). The horizontal arrows on the y-nullcline point left for $x > 0$ and right for $x < 0$ (because $x' = -x$). See Figure 10-1. Orbits coming in from the right or left appear to be asymptotic to the y-axis, some going towards $+\infty$ and others to $-\infty$. See Figure 10-2.

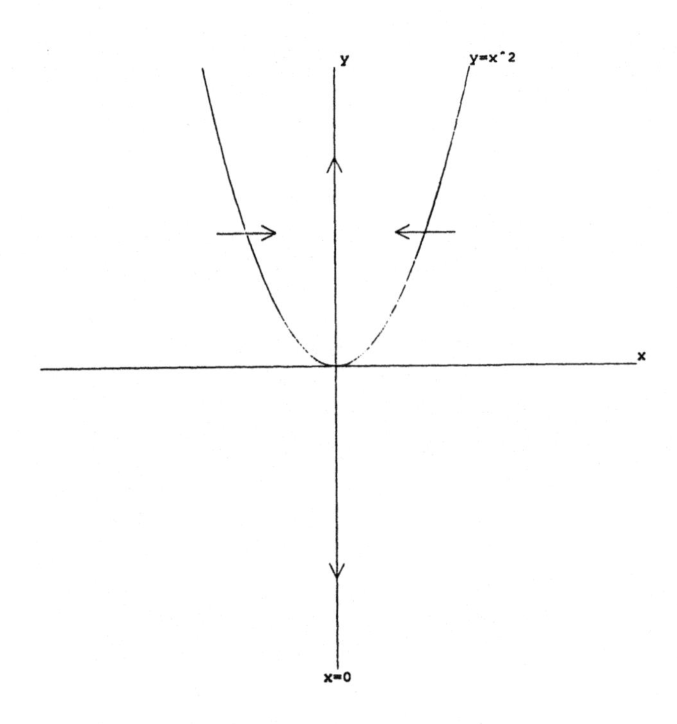

Figure 10-1 Exercise 3. Nullclines.

(f) $x \to 0$ and $y \to \pm\infty$, except that there appears to be an orbit from the right that approaches $(0, 0)$, and a similar one from the left.

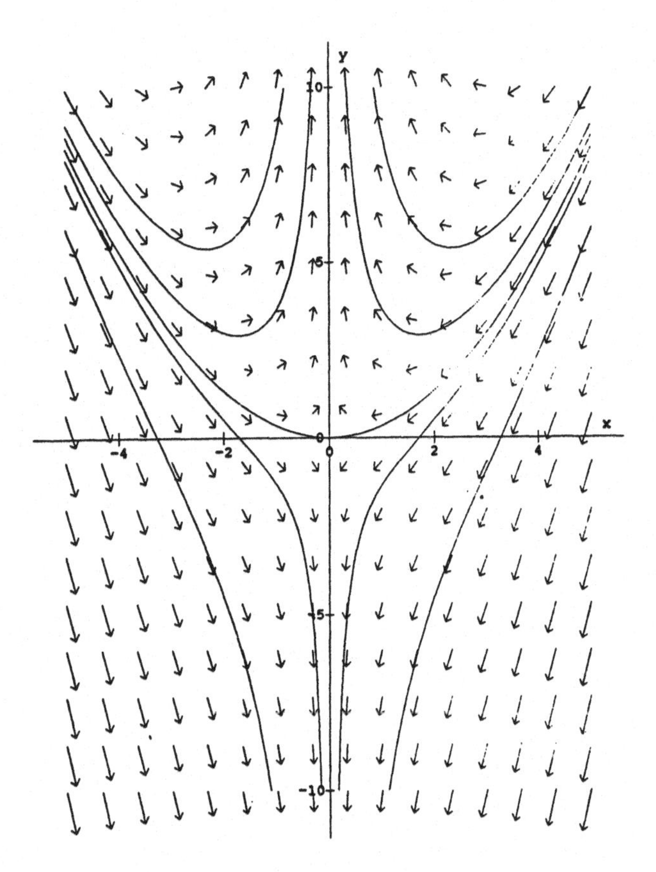

Figure 10-2 Exercise 3. Phase portrait.

(g) There is a separatrix. From part (d) $y = \frac{1}{3}x^2 + C/x$. As $x \to 0$, (i) $y \to \infty$ if $C > 0$, (ii) $y \to -\infty$ if $C < 0$, (iii) $y \to 0$ if $C = 0$ So, $y = \frac{1}{3}x^2$ is a separatrix. There is no basin of attraction, since the equilibrium point $(0, 0)$ does not attract all orbits in its vicinity.

5. **(a)** Constant solutions will satisfy $y = 0$, $2x - x^2 = 0$, that is, $y = 0$, and $x = 0$ or $x = 2$. There are two equilibrium points $(0, 0)$ and $(2, 0)$.

(b) The orbits in the phase plane satisfy

$$\frac{dy}{dx} = \frac{\frac{dy}{dt}}{\frac{dx}{dt}} = \frac{2x - x^2}{y},$$

a separable differential equation. There are no equilibrium solutions. The nonequilibrium solutions are obtained from

$\int y\,dy = \int \left(2x - x^2\right) dx$, that is,

$y^2 = 2x^2 - \frac{2}{3}x^3 + C$. These are the orbits.

(c) The x-nullcline is $y = 0$, and the y-nullcline is $y = 2x - x^2$. The vertical arrows on the x-nullcline point up for $0 < x < 2$ down otherwise (because $y' = x(2 - x)$). The horizontal arrows on the y-nullcline point right for $y > 0$ and left for $y < 0$ (because $x' = y$). See Figure 10-3.

Orbits are given by $y = \pm\sqrt{2x^2 - \frac{2}{3}x^3 + C}$, which are symmetric across the x-axis — so if we can plot $y = \sqrt{2x^2 - \frac{2}{3}x^3 + C}$ the rest of the curve is obtained by flipping this curve across the x-axis. The cubic $2x^2 - \frac{2}{3}x^3 + C$ comes from $+\infty$ and goes to $-\infty$ as x increases. It has three real roots if $C < 0$, so that $2x^2 - \frac{2}{3}x^3 + C$ is negative between the first and second roots, and negative after the third root. Thus, the function $\sqrt{2x^2 - \frac{2}{3}x^3 + C}$ is undefined between the first and second roots, and after the third root. Thus, for $C < 0$, $y = \pm\sqrt{2x^2 - \frac{2}{3}x^3 + C}$ consists of two curves, one that stretches from $-\infty$ and goes to $+\infty$, crosses the x-axis vertically at the first root, and one that is closed and crosses the x-axis vertically at the second and third roots, centered around $(2,0)$. When $C = 0$, the cubic's first and second roots coincide, so the vertical and closed curves for the $C < 0$ case just touch at the first root. When $C > 0$ there is only one real root, and $y = \pm\sqrt{2x^2 - \frac{2}{3}x^3 + C}$ gives a curve similar to the $C = 0$ case except it does not touch the axis at the first root. See Figure 10-4.

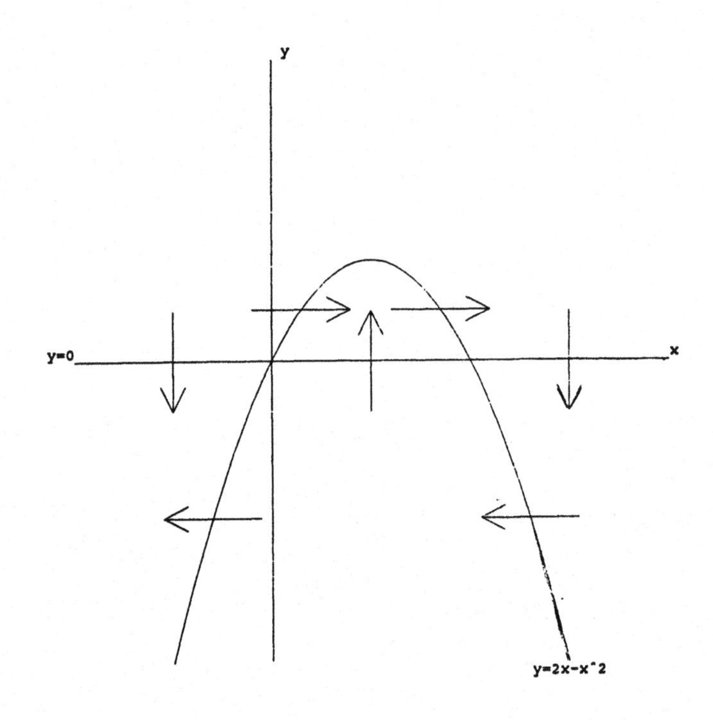

Figure 10-3 Exercise 5. Nullclines.

(d) As x and y increase different behaviors are exhibited. Orbits near $(2,0)$ rotate around $(2,0)$. Almost all other orbits have the property that $x \to -\infty$, and $y \to -\infty$. The orbit described by $C = 0$, that is, $y = \pm\sqrt{2x^2 - \frac{2}{3}x^3}$ approaches $(0,0)$ as t increases.

(e) There is a separatrix. From parts (c) and (d), the orbits $y^2 = 2x^2 - \frac{2}{3}x^3 + C$ behave differently depending on whether C is positive or negative. Thus, when $C = 0$, that is, $y^2 = 2x^2 - \frac{2}{3}x^3$, we have a separatrix. There is no basin of attraction for the equilibrium point $(0,0)$ because it does not attract all orbits in its vicinity. The equilibrium point $(2,0)$ does not attract curves that lie inside the separatrix, so there is no basin of attraction for $(2,0)$.

7. **(a)** Constant solutions will satisfy $x(x - 2y) = 0$, $y(y - 2x) = 0$, that is, $x = 0$ or $x = 2y$, and $y = 0$ or $y = 2x$. There is one equilibrium point, $(0,0)$.

(b) The orbits in the phase plane satisfy

$$\frac{dy}{dx} = \frac{\frac{dy}{dt}}{\frac{dx}{dt}} = \frac{y(y - 2x)}{x(x - 2y)},$$

a differential equation with homogeneous coefficients. Making the change of variables $y = zx$ we find

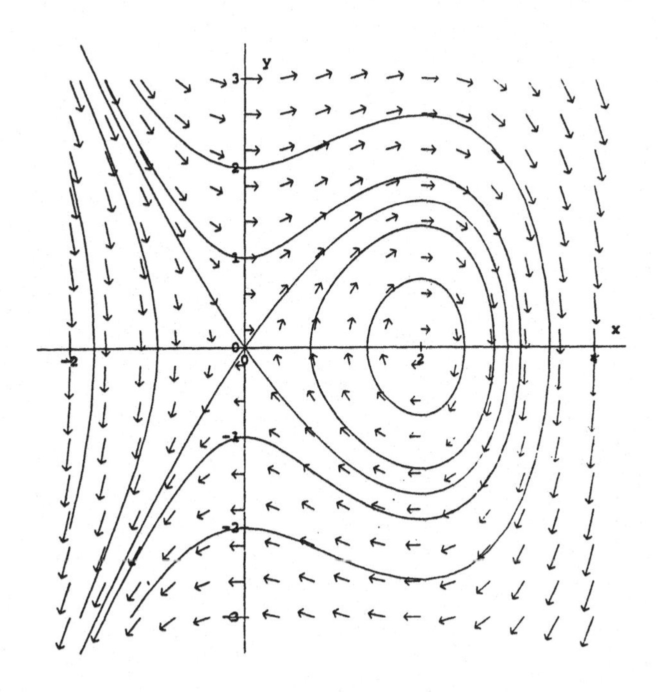

Figure 10-4 Exercise 5. Phase portrait.

$z'x + z = z\,(z-2)\,/\,(1-2z)$, or

$z'x = 3\,(z^2 - z)\,/\,(1-2z)$.

The equilibrium solutions are $z = 0$ and $z = 1$, that is, $y = 0$ and $y = x$ are orbits.

The nonequilibrium solutions are obtained from

$$\int \frac{1-2z}{z^2-z}\,dz = 3\int \frac{1}{x}\,dx.$$

Because $\frac{d}{dz}\left(z^2 - z\right) = -\left(1-2z\right)$, we find $-\ln\left|z^2 - z\right| = 3\ln|x| + C$, or

$z^2 - z = cx^{-3}$.

Substituting $z = y/x$ gives

$y^2 - xy - c/x = 0$, or

$y = \frac{1}{2}\left(x \pm \sqrt{x^2 - 4c/x}\right)$, as the remaining orbits.

(c) The x-nullclines are $x = 0$ and $y = x/2$, and the y-nullclines are $y = 0$ and $y = 2x$. The vertical arrows on the x-nullcline $x = 0$ point up (because, when $x = 0$, $y' = y^2$). The vertical arrows on the x-nullcline $y = x/2$ point down (because, when $y = x/2$, $y' = -\frac{3}{4}x^2$). The horizontal arrows on the y-nullcline $y = 0$ point right (because, when $y = 0$, $x' = x^2$). The horizontal arrows on the y-nullcline $y = 2x$ point left (because, when $y = 2x$, $x' = -3x^2$). See Figure 10-5. Orbits are given by $y = 0$, $y = x$, and $y = \frac{1}{2}\left(x \pm \sqrt{x^2 - 4c/x}\right)$. See Figure 10-6.

(d) Orbits are (i) asymptotic to $x = 0$ for $y > 0$ and $y > x$, (ii) asymptotic to $y = 0$ for $x > 0$ and $y < x$, (iii) asymptotic to $y = x$ for $x < 0$ and $y < 0$, or (iv) approach $(0,0)$ along $y = 0$ for $x < 0$, along $x = 0$ for $y < 0$, and along $y = x$ for $x > 0$.

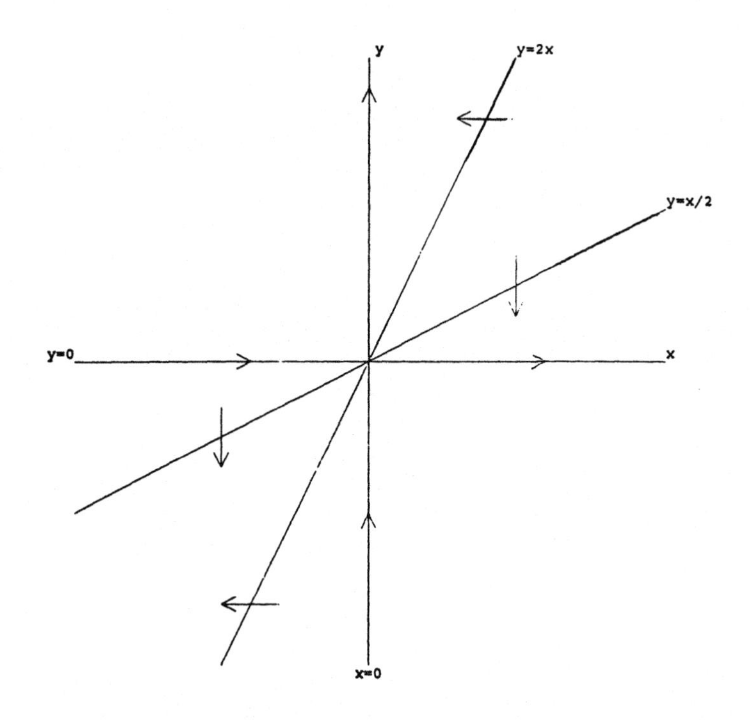

Figure 10-5 Exercise 7. Nullclines.

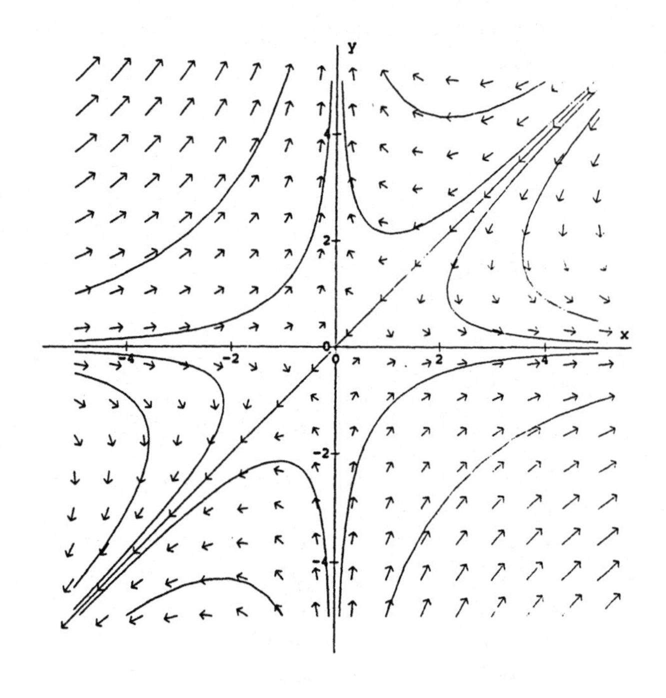

Figure 10-6 Exercise 7. Phase portrait.

(e) There are three separatrices, namely, (i) $y = 0$ for $x < 0$, (ii) $x = 0$ for $y < 0$, and (iii) $y = x$ for $x > 0$. There is no basin of attraction for the equilibrium point $(0, 0)$ because it does not attract all orbits in its vicinity.

9. (a) i. A bead placed on a parabolic shaped wire will oscillate backwards and forwards. Because there is no friction, the maximum height will be the same on both arms of the wire. The phase portrait will consist of concentric closed orbits directed clockwise centered around the origin — the equilibrium point corresponding to the initial condition $x = 0$, $y = 0$, that is, the bead at rest at the minimum for all time. There is no separatrix, but the entire xy-plane is the basin of attraction. See Figure 10-7.

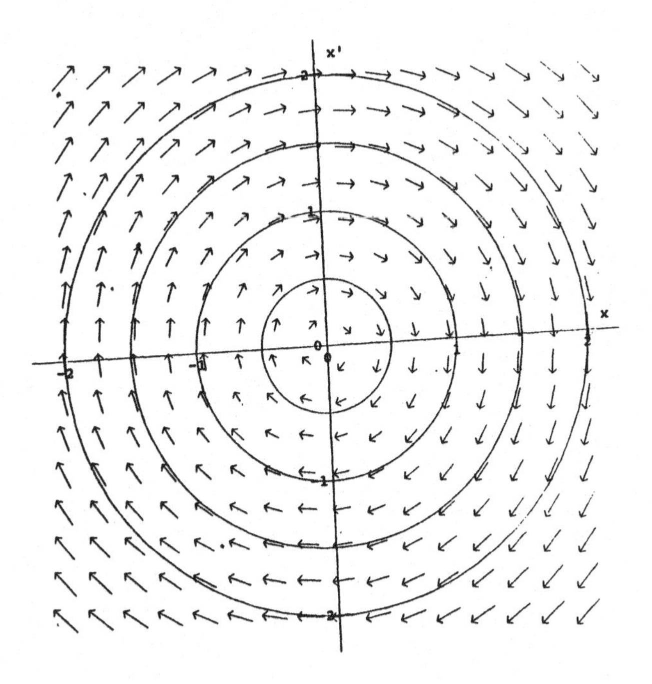

Figure 10-7 Exercise 9(a) (i). Phase portrait.

iii. A bead on a wire shaped like $-x^2(x-1)$ will have two equilibrium points, $x = 0$, $y = 0$, (when the bead at rest at the local minimum for all time) and $x = \frac{1}{2}$, $y = 0$, (when the bead at rest at the local maximum for all time).

A bead starting near minimum will oscillate about the minimum giving concentric closed orbits directed clockwise about $(0, 0)$.

A bead might just have enough energy to reach the equilibrium point $\left(\frac{1}{2}, 0\right)$. In the phase plane this would appear like a closed orbit directed clockwise about $(0, 0)$, starting and ending at $\left(\frac{1}{2}, 0\right)$.

A bead with enough energy to pass over the local maximum would then have $x \to \infty$ and $y \to -\infty$ as t increases, which in the phase plane would have $x \to \infty$, $x' \to \infty$.

A bead starting to the right of the local maximum could be driven towards the local maximum with just enough energy to reach the maximum. In the phase plane this would be represented by a curve approaching $\left(\frac{1}{2}, 0\right)$ from $x > \frac{1}{2}$ for $x' < 0$. Putting this together gives a phase plane similar to Exercise 5, flipped across the y-axis. See Figure 10-8.

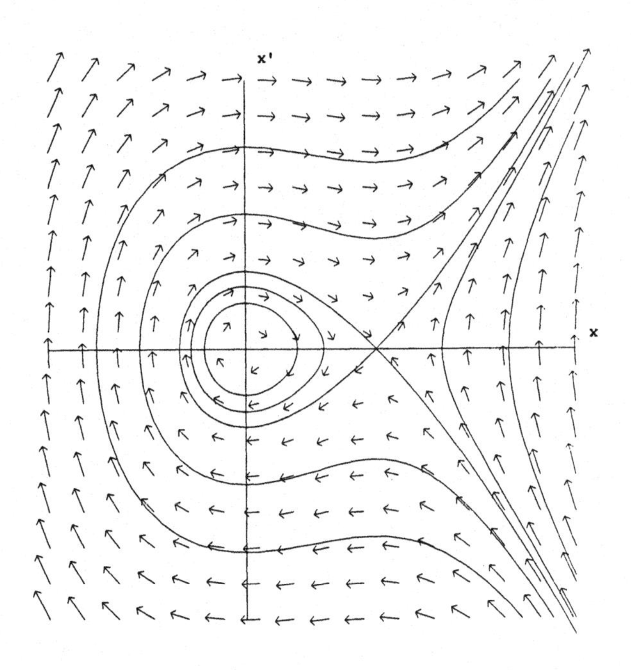

Figure 10-8 Exercise 9(a) (iii). Phase portrait.

10.2 Qualitative Behavior Using Nullclines

1. **(a)** The equilibrium points satisfy the set of equations $x + y = 0$ and $x^2 + y = 0$, which has two solutions, $(0,0)$ and $(1,-1)$. The x-nullcline is $y = -x$ and the y-nullcline is $y = -x^2$. They intersect at the equilibrium points. The vertical arrows on the x-nullcline point up when $x < 0$ or $x > 1$ and down when $0 < x < 1$ (because, when $y = -x$, $y' = x\,(x-1)$). The horizontal arrows on the y-nullcline point right when $0 < x < 1$ and left when $x < 0$ or $x > 1$ (because, when $y = -x^2$, $x' = x\,(1-x)$). Thus, the equilibrium point $(0,0)$ behaves like an unstable focus, while $(1,-1)$ behaves like a saddle point. A computer drawn phase plot is consistent with this information. See Figures 10-9 and 10-10.

(c) The equilibrium points satisfy the set of equations $-y+x^2-1 = 0$ and $-y-x^2+1 = 0$, which has two solutions, $(-1,0)$ and $(1,0)$. The x-nullcline is $y = x^2 - 1$ and the y-nullcline is $y = -x^2 + 1$. They intersect at the equilibrium points. The vertical arrows on the x-nullcline point up when $-1 < x < 1$ and down when $x < -1$ or $x > 1$ (because, when $y = x^2 - 1$, $y' = -2\left(x^2 - 1\right)$). The horizontal arrows on the y-nullcline point left when $-1 < x < 1$ and right when $x < -1$ or $x > 1$ (because, when $y = -x^2 + 1$, $x' = 2\left(x^2 - 1\right)$). The arrows near the equilibrium point $(-1,0)$ all rotate in the same direction (counterclockwise) so the nullcline analysis is inconclusive as far as $(-1,0)$ is concerned — it could be a node, a center, or a focus. The equilibrium point $(1,0)$ behaves like a saddle point. A computer drawn phase plot suggests that $(-1,0)$ behaves like a stable focus, and confirms that $(1,0)$ behaves like a saddle point. See Figures 10-11 and 10-12.

3. **(a)** The equilibrium points satisfy the set of equations $-x = 0$ and $y - x^2 = 0$, which

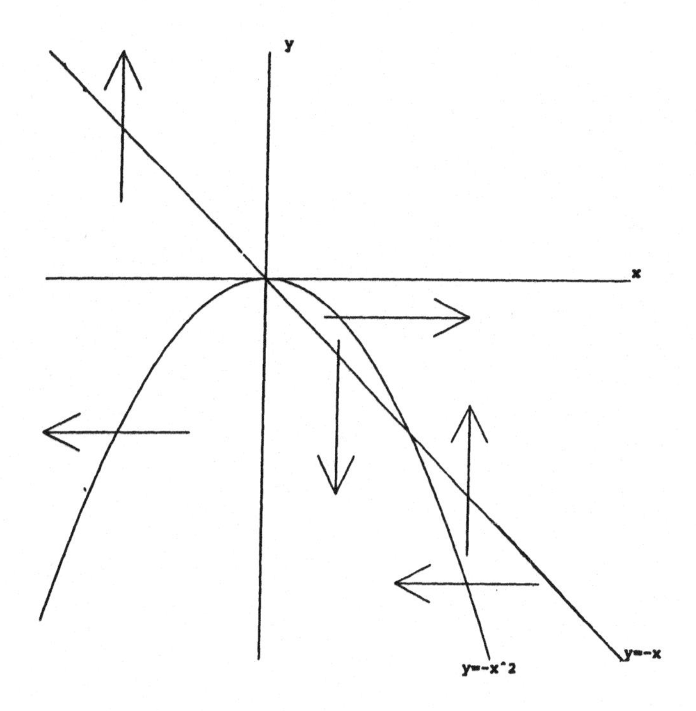

Figure 10-9 Exercise 1(a). Nullclines.

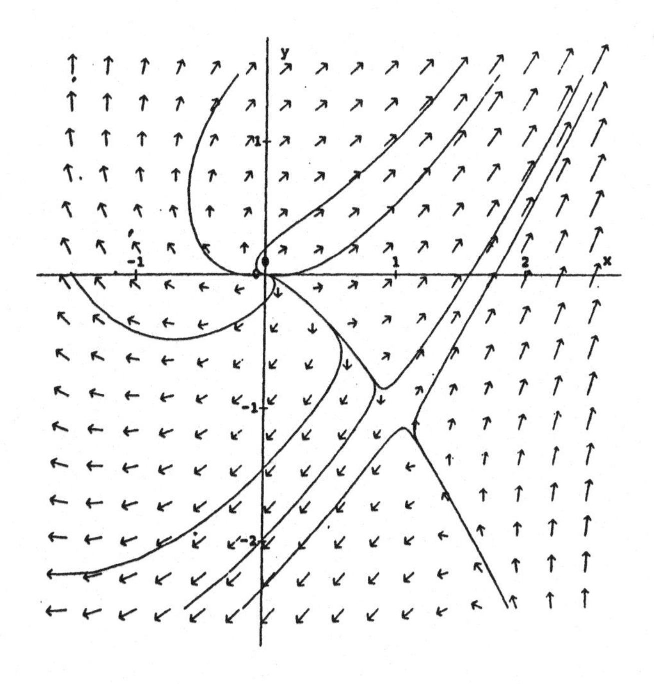

Figure 10-10 Exercise 1(a). Phase portrait.

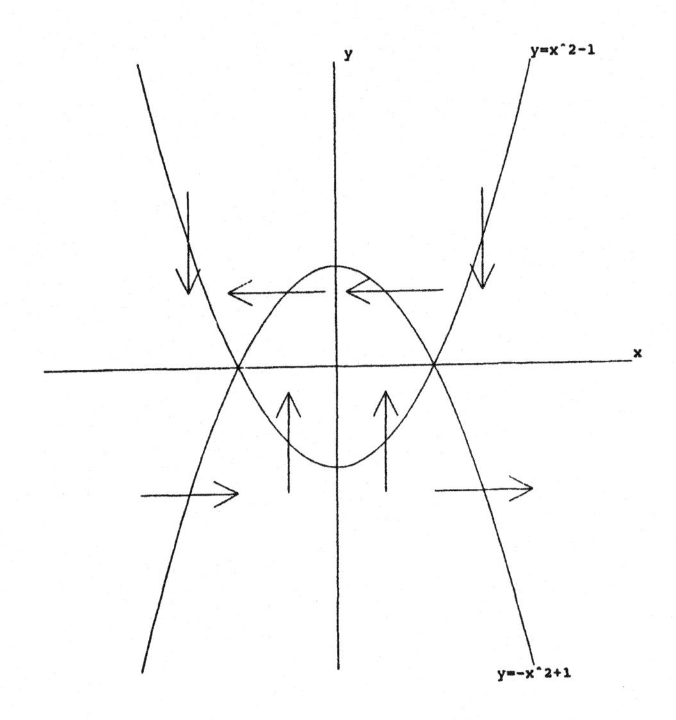

Figure 10-11 Exercise 1(c). Nullclines.

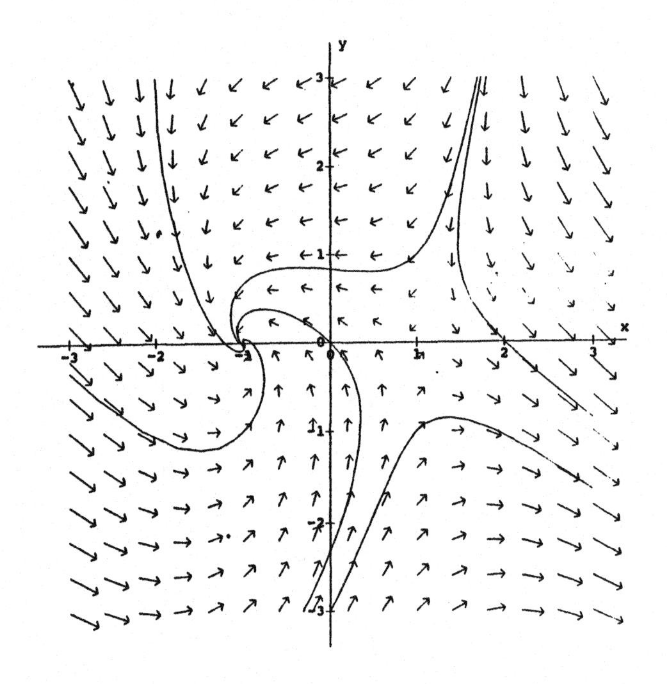

Figure 10-12 Exercise 1(c). Phase portrait.

has the solution $(0,0)$. The x-nullcline is $x = 0$, and the y-nullcline is $y = x^2$. They intersect at the equilibrium point. The vertical arrows on the x-nullcline point up for $y > 0$ and down for $y < 0$ (because, when $x = 0$, $y' = y$). The horizontal arrows on the y-nullcline point left for $x > 0$ and right for $x < 0$ (because $x' = -x$). Thus, the equilibrium point $(0,0)$ behaves like a saddle point. See Figure 10-2.

(c) The equilibrium points satisfy the set of equations $y = 0$, $2x - x^2 = 0$. There are two equilibrium points, $(0,0)$ and $(2,0)$. The x-nullcline is $y = 0$, and the y-nullcline is $y = 2x - x^2$. They intersect at the equilibrium points. The vertical arrows on the x-nullcline point up for $0 < x < 2$ down otherwise (because $y' = x(2 - x)$). The horizontal arrows on the y-nullcline point right for $y > 0$ and left for $y < 0$ (because $x' = y$). The arrows near the equilibrium point $(2,0)$ all rotate in the same direction (clockwise) so the nullcline analysis is inconclusive as far as $(2,0)$ is concerned — it could be a node, a center, or a focus. The equilibrium point $(0,0)$ behaves like a saddle point. A computer drawn phase plot suggests that $(2,0)$ behaves like a center, and confirms that $(0,0)$ behaves like a saddle point. See Figure 10-4.

(e) The equilibrium points satisfy the set of equations $x(x - 2y) = 0$, $y(y - 2x) = 0$. There is one equilibrium point, $(0,0)$. The x-nullclines are $x = 0$ and $y = x/2$, and the y-nullclines are $y = 0$ and $y = 2x$. They intersect at the equilibrium point. The vertical arrows on the x-nullcline $x = 0$ point up (because, when $x = 0$, $y' = y^2$). The vertical arrows on the x-nullcline $y = x/2$ point down (because, when $y = x/2$, $y' = -\frac{3}{4}x^2$). The horizontal arrows on the y-nullcline $y = 0$ point right (because, when $y = 0$, $x' = x^2$). The horizontal arrows on the y-nullcline $y = 2x$ point left (because, when $y = 2x$, $x' = -3x^2$). The equilibrium point $(0,0)$ behaves like a saddle point, except that there are three orbits that are attracted to the equilibrium point — $y = x$ for $x > 0$, $x = 0$ for $y < 0$, and $y = 0$ for $x < 0$ — instead of the usual two. See Figure 10-6.

10.3 Qualitative Behavior Using Linearization

1. (a) Figure 10-13 shows a saddle point behavior at $(-1,0)$, and an unstable node behavior at $(1,0)$.

(c) Figure 10-14 shows saddle point behaviors at $(-1,0)$ and $(1,0)$.

(e) It is not possible to have two stable nodes without a third equilibrium point — which, for example, behaves like a saddle point — between them.

3. (a) The equilibrium points are $(0,0)$ and $(1,-1)$. Here $P(x,y) = x + y$ and $Q(x,y) = x^2 + y$, so $P_x = 1$, $P_y = 1$, $Q_x = 2x$, and $Q_y = 1$, with Jacobean matrix

$$J(x,y) = \begin{bmatrix} 1 & 1 \\ 2x & 1 \end{bmatrix}.$$

At $(0,0)$ we find

$$J(0,0) = \begin{bmatrix} 1 & 1 \\ 0 & 1 \end{bmatrix},$$

with determinant $D = 1$ and trace $T = 1$, so $\Delta = T^2 - 4D = -3 < 0$, which is characteristic of an unstable focus.

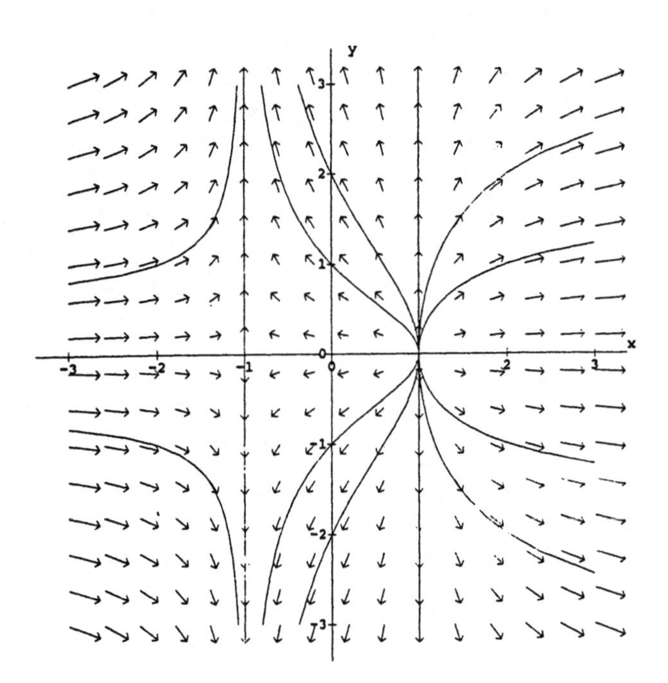

Figure 10-13 Exercise 1(a). Phase portrait.

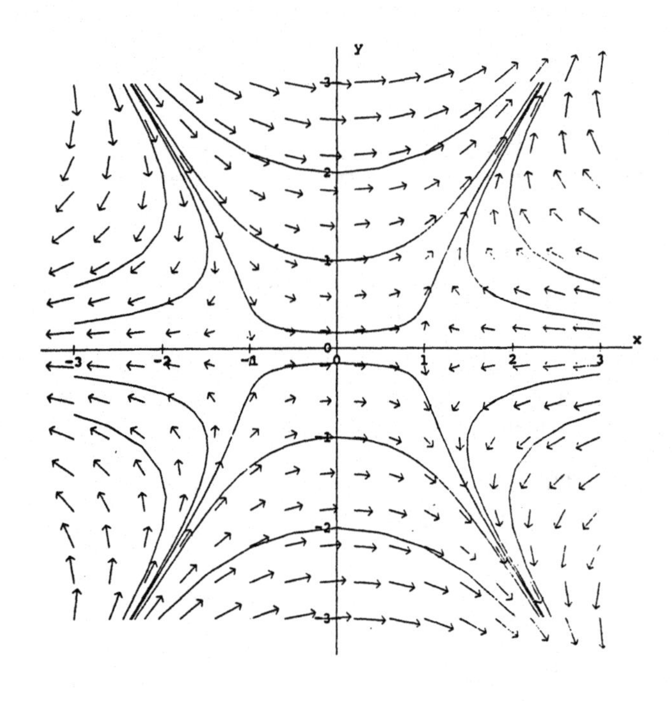

Figure 10-14 Exercise 1(c). Phase portrait.

At $(1,-1)$ we find

$$J(1,-1) = \begin{bmatrix} 1 & 1 \\ 2 & 1 \end{bmatrix},$$

with determinant $D = -1$ and trace $T = 1$, so $\Delta = T^2 - 4D = 5 > 0$, which is characteristic of a saddle point. See Figure 10-10.

(c) The equilibrium points are $(-1,0)$ and $(1,0)$. Here $P(x,y) = -y + x^2 - 1$ and $Q(x,y) = -y - x^2 + 1$, so $P_x = 2x$, $P_y = -1$, $Q_x = -2x$, and $Q_y = -1$, with Jacobean matrix

$$J(x,y) = \begin{bmatrix} 2x & -1 \\ -2x & -1 \end{bmatrix}.$$

At $(-1,0)$ we find

$$J(-1,0) = \begin{bmatrix} -2 & -1 \\ 2 & -1 \end{bmatrix},$$

with determinant $D = 4$ and trace $T = -3$, so $\Delta = T^2 - 4D = -7 < 0$, which is characteristic of a stable focus.

At $(1,0)$ we find

$$J(1,0) = \begin{bmatrix} 2 & -1 \\ -2 & -1 \end{bmatrix},$$

with determinant $D = -4$ and trace $T = 1$, so $\Delta = T^2 - 4D = 17 > 0$, which is characteristic of a saddle point. See Figure 10-12.

5. (a) The equilibrium point is $(0,0)$. Here $P(x,y) = -x$ and $Q(x,y) = y - x^2$, so $P_x = -1$, $P_y = 0$, $Q_x = -2x$, and $Q_y = 1$, with Jacobean matrix

$$J(x,y) = \begin{bmatrix} -1 & 0 \\ -2x & 1 \end{bmatrix}.$$

At $(0,0)$ we find

$$J(0,0) = \begin{bmatrix} -1 & 0 \\ 0 & 1 \end{bmatrix},$$

with determinant $D = -1$ and trace $T = 0$, so $\Delta = T^2 - 4D = 4 > 0$, which is characteristic of a saddle point. See Figure 10-2.

(c) The equilibrium points are $(0,0)$ and $(2,0)$. Here $P(x,y) = y$ and $Q(x,y) = 2x - x^2$, so $P_x = 0$, $P_y = 1$, $Q_x = 2 - 2x$, and $Q_y = 0$, with Jacobean matrix

$$J(x,y) = \begin{bmatrix} 0 & 1 \\ 2 - 2x & 0 \end{bmatrix}.$$

At $(0,0)$ we find

$$J(0,0) = \begin{bmatrix} 0 & 1 \\ 2 & 0 \end{bmatrix},$$

with determinant $D = -2$ and trace $T = 0$, so $\Delta = T^2 - 4D = 8 > 0$, which is characteristic of a saddle point.

At $(2,0)$ we find

$$J(1,0) = \begin{bmatrix} 0 & 1 \\ -2 & 0 \end{bmatrix},$$

with determinant $D = 2$ and trace $T = 0$, so $\Delta = T^2 - 4D = -8 < 0$, which is inconclusive — it could be a center, a stable focus, or an unstable focus. A computer drawn phase plot suggests that $(2,0)$ behaves like a center, and confirms that $(0,0)$ behaves like a saddle point. See Figure 10-4.

(e) The equilibrium point is $(0,0)$. Here $P(x,y) = x(x-2y)$ and $Q(x,y) = y(y-2x)$, so $P_x = 2x - 2y$, $P_y = -2x$, $Q_x = 2y - 2x$, and $Q_y = 2y$, with Jacobean matrix

$$J(x,y) = \begin{bmatrix} 2x-2y & -2x \\ 2y-2x & 2y \end{bmatrix}.$$

At $(0,0)$ we find

$$J(0,0) = \begin{bmatrix} 0 & 0 \\ 0 & 0 \end{bmatrix},$$

with determinant $D = 0$ and trace $T = 0$, so $\Delta = T^2 - 4D = 0$, which is inconclusive. See Figure 10-6.

10.4 Models Involving Nonlinear Autonomous Equations

1. (a) The equilibrium points will satisfy $y = 0$ and $x + 2bx^3 = x(1 + 2bx^2) = 0$, where $b > 0$ for a hard spring, and $b < 0$ for a soft. If the spring is hard, then $x(1+2bx^2) = 0$ has one solution, $x = 0$, so the only equilibrium point is $(0,0)$. If the spring is soft, then $x(1+2bx^2) = 0$ has three solutions, $x = 0$, and $x = \pm 1/\sqrt{-2b}$, so the equilibrium points are $(0,0)$, $(1/\sqrt{-2b}, 0)$, and $(-1/\sqrt{-2b}, 0)$.

(b) For the hard spring, the x-nullcline is $y = 0$, and the y-nullcline is $x = 0$. They intersect at the equilibrium point. The vertical arrows on the x-nullcline point down for $0 < x$ and up for $x < 0$. The horizontal arrows on the y-nullcline point right for $y > 0$ and left for $y < 0$. The arrows all rotate in the same direction (clockwise) so the nullcline analysis is inconclusive — the equilibrium point $(0,0)$ could be a node, a center, or a focus.

For the soft spring, the x-nullcline is $y = 0$, and the y-nullclines are $x = 0$ and $x = \pm 1/\sqrt{-2b}$. They intersect at the equilibrium points. The vertical arrows on the x-nullcline point down for $x < -1/\sqrt{-2b}$ and $0 < x < 1/\sqrt{-2b}$, and up for $-1/\sqrt{-2b} < x < 0$ and $1/\sqrt{-2b} < x$. The horizontal arrows on the y-nullcline point right for $y > 0$ and left for $y < 0$. The equilibrium points $(1/\sqrt{-2b}, 0)$, and $(-1/\sqrt{-2b}, 0)$ behave like saddle points, while near the equilibrium point $(0,0)$, the arrows all rotate in the same direction (clockwise), so the nullcline analysis is inconclusive in this case — $(0,0)$ could be a node, a center, or a focus.

(c) Here $P(x,y) = y$ and $Q(x,y) = -\lambda^2(x + 2bx^3)$, so $P_x = 0$, $P_y = 1$, $Q_x = -\lambda^2(1 + 6bx^2)$, and $Q_y = 0$, with Jacobean matrix

$$J(x,y) = \begin{bmatrix} 0 & 1 \\ -\lambda^2(1+6bx^2) & 0 \end{bmatrix}.$$

At $(0,0)$ we find

$$J(0,0) = \begin{bmatrix} 0 & 1 \\ -\lambda^2 & 0 \end{bmatrix},$$

with determinant $D = \lambda^2$ and trace $T = 0$, so $\Delta = T^2 - 4D = -4\lambda^2 < 0$, which is inconclusive, for both the hard and soft springs.

For the soft spring, at $(\pm 1/\sqrt{-2b}, 0)$ we find

$$J\left(\pm 1/\sqrt{-2b}, 0\right) = \begin{bmatrix} 0 & 1 \\ -2\lambda^2 & 0 \end{bmatrix},$$

with determinant $D = -2\lambda^2$ and trace $T = 0$, so $\Delta = T^2 - 4D = 8\lambda^2 > 0$, which characterize saddle points.

(d) The orbits in the phase plane satisfy

$$\frac{dy}{dx} = \frac{\frac{dy}{dt}}{\frac{dx}{dt}} = -\lambda^2 \frac{x + 2bx^3}{y},$$

a separable differential equation. There are no equilibrium solutions. The nonequilibrium solutions are obtained from
$\int y\,dy = -\lambda^2 \int (x + 2bx^3)\,dx$, that is,
$y^2 = -\lambda^2 (x^2 + bx^4) + C$. These are the orbits.

(e) Figure 10.37 in the text is a phase portrait for a hard spring, because it has only one equilibrium point, and agrees with parts (b) and (c). Figure 10.38 in the text is a phase portrait for a soft spring, because it has three equilibrium points, and agrees with parts (b) and (c).

(f) The motion is periodic.

(g) The closed orbits represent periodic motion. This is realistic. Other orbits tell us that, under certain initial conditions, the spring is not strong enough to return to the equilibrium position, but continues stretching forever. This is realistic. Other orbits tell as that, under certain initial conditions, the spring is not strong enough to return to the equilibrium position, but continues contracting forever. This is not realistic.

(h) If $x = x_0$, $y = 0$, then $y^2 = -\lambda^2 (x^2 + bx^4) + C$ gives $C = \lambda^2 (x_0^2 + bx_0^4)$, and so $y^2 = \lambda^2 \left[x_0^2 - x^2 + b(x_0^4 - x^4)\right]$. For closed orbits, the period T is 4 times the time it takes for the spring starting from rest at x_0 to reach $x = 0$. Because $y = x'$ we have $x' = -\lambda\sqrt{x_0^2 - x^2 + b(x_0^4 - x^4)}$, where the minus sign is selected because initially the velocity is negative. Thus, integrating this from $t = 0$ and $x = x_0$, to $t = T/4$ and $x = 0$ gives

$$\int_{x_0}^{0} \frac{1}{\lambda\sqrt{x_0^2 - x^2 + b(x_0^4 - x^4)}}\,dx = -\int_0^{T/4} dt,$$

or, using $\int_{x_0}^{0} f(x)\,dx = -\int_0^{x_0} f(x)\,dx$,

$$T = \frac{4}{\lambda}\int_0^{x_0} \frac{1}{\sqrt{x_0^2 - x^2 + b(x_0^4 - x^4)}}\,dx.$$

Making the substitution $x = x_0 \sin u$, so

$$\begin{aligned} x_0^2 - x^2 + b(x_0^4 - x^4) &= (x_0^2 - x^2)\left[1 + b(x_0^2 + x^2)\right] \\ &= (x_0^2 - x_0^2\sin^2 u)\left[1 + b(x_0^2 + x_0^2\sin^2 u)\right] \\ &= x_0^2\cos^2 u\left[1 + b(x_0^2 + x_0^2\sin^2 u)\right], \end{aligned}$$

and $dx = x_0 \cos u\,du$, while $x = 0$ corresponds to $u = 0$ and $x = x_0$ corresponds to $u = \pi/2$, we find

$$T = \frac{4}{\lambda}\int_0^{\pi/2} \frac{1}{\sqrt{1 + b(x_0^2 + x_0^2\sin^2 u)}}\,dx.$$

For this integral to exist we require $1 + b(x_0^2 + x_0^2\sin^2 u) \geq 0$ for $0 \leq u \leq \pi/2$. The largest value that the left-hand side can attain is at $u = \pi/2$ in which case $\sin u = 1$ and $1 + b(x_0^2 + x_0^2\sin^2 u) = 1 + 2bx_0^2$. If $b > 0$ this is always positive. If $b < 0$ then for $1 + 2bx_0^2$ to be positive we need $2bx_0^2 \geq -1$, that is, $x_0^2 \leq -1/(2b)$. In Figure 10.37 of the text we have $b = \frac{1}{2}$, so solutions will be periodic for any initial x_0. In Figure 10.38 of the text we have $b = -\frac{1}{2}$, so solutions will be periodic if $x_0^2 \leq 1$, which they are. For a soft spring it takes longer for the outer closed curves to complete a period than the inner. For a hard spring it takes longer for the inner closed curves to complete a period than the outer.

3. **(a)** The equilibrium point is $(0,0)$.

(b) See **Figure** 10-15. All orbits seem to be approaching a limit cycle with "center" $(0,0)$ and "radius" approximately 2.

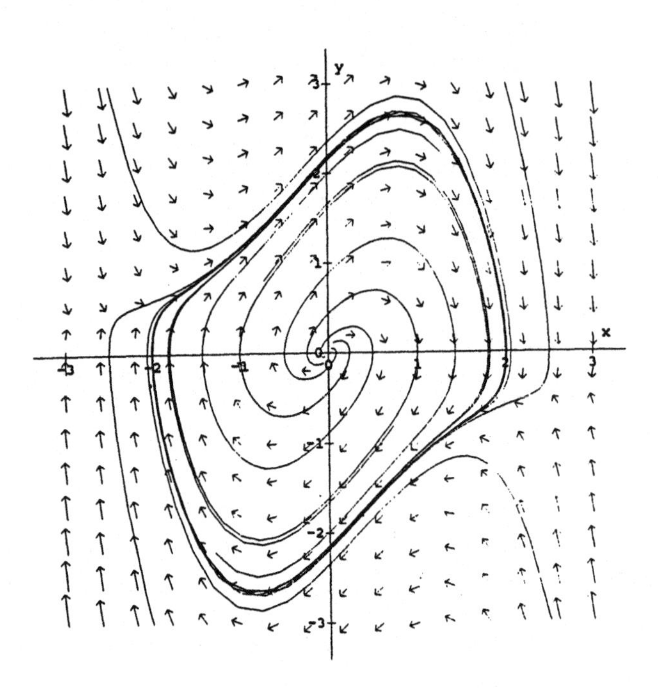

Figure 10-15 Exercise 3. Phase portrait.

(c) The x-nullcline is $y = 0$, and the y-nullcline is $y = x/\left(x^2 - 1\right)$. They intersect at the equilibrium point. The vertical arrows on the x-nullcline point up for $y < 0$ and down for $0 < y$ (because, when $x = 0$, $y' = -y$). The horizontal arrows on the y-nullcline point right for $y > 0$ and left for $y < 0$ (because $x' = -y$). Because the arrows all rotate in the same direction (clockwise) near the equilibrium point, the nullcline analysis is inconclusive — the equilibrium point $(0,0)$ could be a node, a center, or a focus.

(d) The equilibrium point is $(0,0)$. Here $P(x,y) = y$ and $Q(x,y) = -x - \left(x^2 - 1\right)y$, so $P_x = 0$, $P_y = 1$, $Q_x = -1 - 2xy$, and $Q_y = -\left(x^2 - 1\right)$, with Jacobean matrix

$$J(x,y) = \begin{bmatrix} 0 & 1 \\ -1 - 2xy & -\left(x^2 - 1\right) \end{bmatrix}.$$

At $(0,0)$ we find

$$J(0,0) = \begin{bmatrix} 0 & 1 \\ -1 & 1 \end{bmatrix},$$

with determinant $D = 1$ and trace $T = 1$, so $\Delta = T^2 - 4D = -3$. This means that $(0,0)$ behaves like an unstable focus, which accounts for the local spiralling of orbits away from the origin.

(e) The differential equation for circular orbits is $dy/dx = -x/y$. Thus, if $x^2 - 1 > 0$, then the equation $dy/dx = -x/y - \left(x^2 - 1\right)$ will have slopes more negative than that of the slopes for a circle at the same point. This means that whenever if $x^2 - 1 > 0$ orbits will move towards the origin. The condition $x^2 - 1 > 0$ is equivalent to the conditions $x > 1$ and $x > -1$, that is, $|x| > 1$. A similar argument applies when $x^2 - 1 < 0$ to show that whenever an orbit has $|x| < 1$ it moves away from the origin.

(f) Consider an orbit that starts with $|x| > 1$. It will be attracted to the origin. However, if it passes the vertical lines $x = \pm 1$, then it will be repelled from the origin. This is consistent with the limit cycle observed in part (b).

5. Because x and y represent populations, we are concerned only with $x \geq 0$ and $y \geq 0$. The equilibrium points satisfy the set of equations $x(9 - x - y) = 0$ and $y(6 - x - y) = 0$, which has the three solutions $(0,0)$, $(0,6)$, and $(9,0)$. We perform both a nullcline analysis and a linearization analysis.

Nullcline Analysis. The x-nullclines are $x = 0$ and $y = 9 - x$. The y-nullclines are $y = 0$, and $y = 6 - x$. They intersect at the equilibrium points. (Note that the two x-nullclines intersect at $(0,9)$ and the two y-nullclines intersect at $(6,0)$. Neither of these are equilibrium points.) The vertical arrows on the x-nullcline $x = 0$ point up for $y < 6$ and down for $y > 6$ (because, when $x = 0$, $y' = y(6 - y)$). The vertical arrows on the x-nullcline $y = 9 - x$ point down (because, when $y = 9 - x$, $y' = -3y$). The horizontal arrows on the y-nullcline $y = 0$ point right for $x < 9$ and left for $x > 9$ (because, when $y = 0$, $x' = x(9 - x)$). The horizontal arrows on the y-nullcline $y = 6 - x$ point right (because, when $y = 6 - x$, $x' = 3x$). Thus, the equilibrium point $(0,0)$ behaves like an unstable node, the equilibrium point $(0,6)$ behaves like a saddle point, and the equilibrium point $(9,0)$ behaves like a stable node. See Figure 10-16. Thus, if a solution does not start at $(0,0)$ or $(0,6)$, all solutions tend to $(9,0)$ — that is, the y populations becomes extinct.

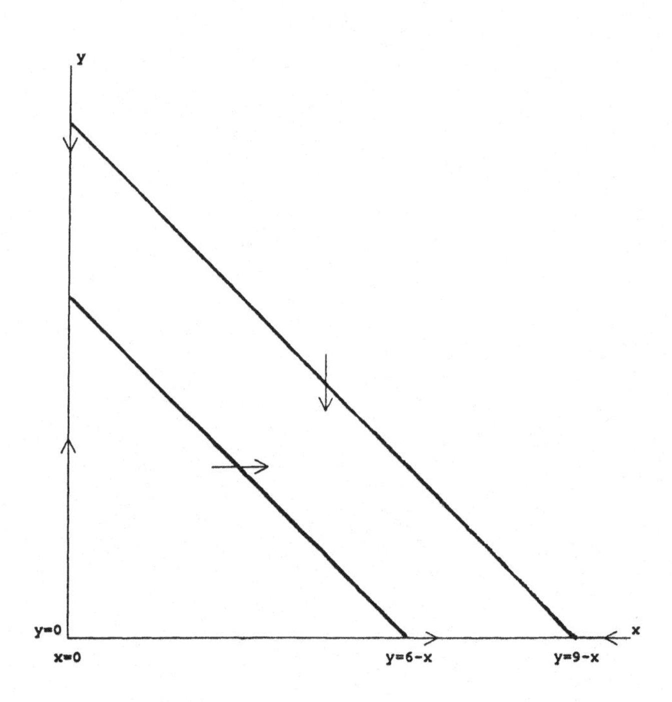

Figure 10-16 Exercise 5. Nullclines.

Linearization Analysis. Here $P(x,y) = x(9 - x - y)$ and $Q(x,y) = y(6 - x - y)$, so Jacobean matrix is

$$J(x,y) = \begin{bmatrix} 9 - 2x - y & -x \\ -y & 6 - x - 2y \end{bmatrix}.$$

At $(0,0)$ we find

$$J(0,0) = \begin{bmatrix} 9 & 0 \\ 0 & 6 \end{bmatrix},$$

with determinant $D = 54$ and trace $T = 15$, so $\Delta = T^2 - 4D = 9$. This means that $(0,0)$ behaves like an unstable node.

At $(0,6)$ we find

$$J(0,0) = \begin{bmatrix} 3 & 0 \\ -6 & -6 \end{bmatrix},$$

with determinant $D = -18$ and trace $T = -3$, so $\Delta = T^2 - 4D = 81$. This means that $(0,0)$ behaves like a saddle point.

At $(9,0)$ we find

$$J(0,0) = \begin{bmatrix} -9 & -9 \\ 0 & -3 \end{bmatrix},$$

with determinant $D = 27$ and trace $T = -12$, so $\Delta = T^2 - 4D = 36$. This means that $(0,0)$ behaves like a stable node.

A phase portrait is shown in Figure 10-17, where there appears to be a straight line orbit joining the equilibrium points $(0,6)$ and $(9,0)$.

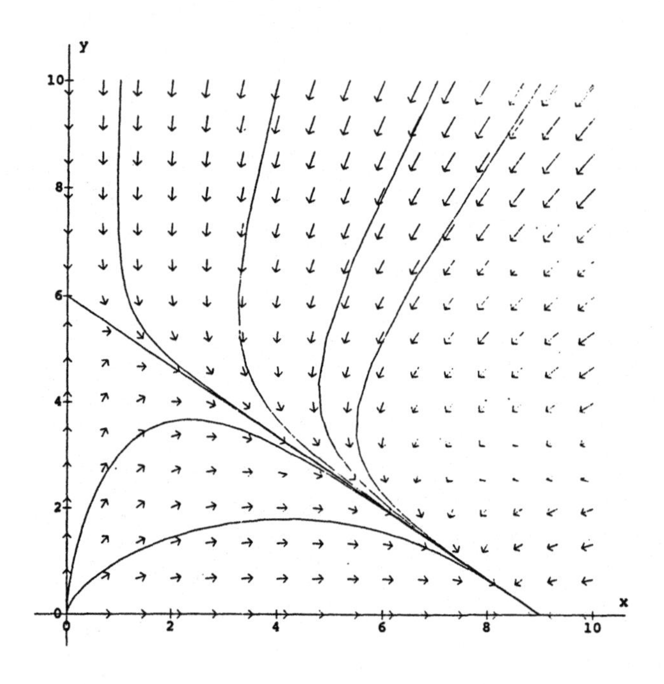

Figure 10-17 Exercise 5. Phase portrait.

Straight Line Orbits. To find all straight line orbits we substitute $y = mx + b$ into $x' = x(9 - x - y)$ and $y' = y(6 - x - y)$ and try to find conditions on m and b. Doing this leads to

$x' = x(9 - x - mx - b) = x[9 - b - (m+1)x]$ and

$mx' = (mx + b)(6 - x - mx - b) = (mx + b)[6 - b - (m+1)x]$.

Eliminating x' from these equations leads to

$mx\,[9 - b - (m+1)\,x] = (mx + b)\,[6 - b - (m+1)\,x]$,

which must be an identity in x. Equating coefficients of x^0, x^1, and x^2 gives

$0 = b\,(6 - b)$, $m\,(9 - b) = m\,(6 - b) - b\,(m+1)$, and $-m\,(m+1) = -m\,(m+1)$.

Solving this system for m and b gives

$m = -2/3$ and $b = 6$. Thus, the only straight line orbit is

$y = -2x/3 + 6$ — the straight line joining $(0, 6)$ and $(9, 0)$.

10.5 Bungee Jumping

1. The conditions on $f(x)$ show that $f(x)$ has the general shape of a parabola opening downwards, crossing the x-axis at x_1 and x_2. This means that $\sqrt{f(x)}$ is defined only for $x_1 \le x \le x_2$. The graph of the function $-\sqrt{f(x)}$ is obtained from that of $\sqrt{f(x)}$ by flipping it across the x-axis, resulting in a closed curve as the graph of $y^2 = f(x)$. If we write $y^{2n} = 1 - x^{2m}$, we see that $1 - x^{2m} = 0$ at $x = \pm 1$, $1 - x^{2m} > 0$ if $-1 < x < 1$, and $1 - x^{2m} < 0$ if $x < -1$, or $x > 1$. But $y^{2n} \ge 0$, so $y = \pm\left(1 - x^{2m}\right)^{1/n}$ has the graph of a closed curve.

3. By rewriting $F(x) = kx - kL^3/x^2$ in the form $x^2 F(x) = kx^3 - kL^3$, and then plotting the data set as $x^2 F(x)$ versus x^3 we find an approximate straight line with slope 2.558 and intercept -432.446. Thus, we can estimate $k \approx 2.558$ and $-kL^3 \approx -432.446$, so that $F(x) \approx 2.558x - 432.446/x^2$. Figure 10-18 shows a plot of the data set together with the function $2.558x - 432.446/x^2$. This is a reasonable fit.

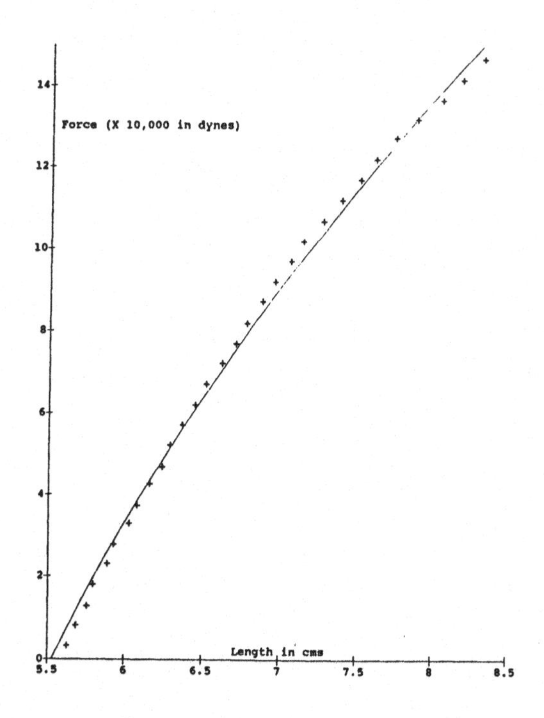

Figure 10-18 Exercise 3.

10.6 Linear Versus Nonlinear Differential Equations

1. The dependent variable x is missing, so we let $y = x'$ and $y' = x''$ in $tx''+x'-4t = 0$ to find $ty'+y-4t = 0$. This is a linear differential equation with solution $y = 2t+C_1/t$. This gives rise to $x' = 2t + C_1/t$, which can be integrated to yield $x(t) = t^2 + C_1 \ln|t| + C_2$.

3. (a) The dependent variable x is missing, so we let $y = x'$ and $y' = x''$ in $2tx''+(x')^2 - 1 = 0$ to find
   $2ty' + y^2 - 1 = 0.$
   This has the equilibrium solutions $y = \pm 1$, that is, $x' = \pm 1$, which gives $x(t) = \pm t + C$.
   The nonequilibrium solutions are obtained from
   $$2\int \frac{1}{y^2-1}\,dy = -\int \frac{1}{t}\,dt,$$
   namely,
   $y = (t + C_1) / (t - C_1)$.
   This gives rise to
   $x' = (t + C_1) / (t - C_1)$, which can be integrated to yield
   $x(t) = t + 2C_1 \ln|t - C_1| + C_2$. Thus, the complete solution is
   $x(t) = \pm t + C$ and $x(t) = t + 2C_1 \ln|t - C_1| + C_2$.

   (c) The independent variable t is missing so we let $y = dx/dt$ and $x'' = ydy/dx$ in $xx'' + (x')^2 = 0$ to find
   $xydy/dx + y^2 = 0$, or
   $xdy/dx + y = 0$.
   This has the equilibrium solution $y = 0$, while the nonequilibrium solutions are obtained from
   $\int dy/y = -\int dx/x$, namely, $\ln|y| = -\ln|x| + C$. Thus,
   $y = C_1/x$ contains both the equilibrium and the nonequilibrium solutions.
   $dx/dt = C_1/x$ can be integrated immediately to find
   $x^2 = 2C_1 t + C_2$.

   (e) The independent variable t is missing so we let $y = dx/dt$ and $x'' = ydy/dx$ in $(x^2+1)\,x'' - 2x\,(x')^2 = 0$ to find
   $(x^2+1)\,ydy/dx - 2xy^2 = 0$, or
   $(x^2+1)\,dy/dx - 2xy = 0$.
   This has the equilibrium solution $y = 0$, while the nonequilibrium solutions are obtained from
   $\int dy/y = \int 2x\,dx/\left(x^2+1\right)$, namely,
   $\ln|y| = \ln\left(x^2+1\right) + C$. Thus,
   $y = C_1\left(x^2+1\right)$ and
   $dx/dt = C_1\left(x^2+1\right)$ can be integrated immediately to find
   $\arctan x = C_1 t + C_2$.

5. We make the substitution $v = y'$, then $y'' = \kappa\left[1+(y')^2\right]^{3/2}$ becomes $v' = \kappa\left(1+v^2\right)^{3/2}$. This gives
   $$\int \frac{dv}{(1+v^2)^{3/2}} = \kappa \int dx.$$

The integral on the left-hand side can be evaluated by making the substitution $v = \tan\theta$ to give $\frac{v}{\sqrt{1+v^2}} = \kappa x + C$. Solving this for v yields

$$v = \frac{\pm(\kappa x + C)}{\sqrt{1-(\kappa x + C)^2}},$$

which, from $v = y'$, gives

$$\frac{dy}{dx} = \frac{\pm(\kappa x + C)}{\sqrt{1-(\kappa x + C)^2}}.$$

Integration leads to $y(x) = \mp\frac{1}{\kappa}\sqrt{1-(\kappa x + C)^2} + a$, where a is the constant of integration. This equation can be rewritten in the form $(y-a)^2 + (x-b)^2 = \frac{1}{\kappa^2}$, where we have put $b = -C/\kappa$. This is a circle with center (b, a) and radius $1/\kappa$. Thus, the circle is the only curve with constant curvature.

7. From $x(t) = e^t$, $x' = e^t$, $x'' = e^t$, so

$xx'' - 2(x')^2 + x^2 = e^t e^t - 2(e^t)^2 + (e^t)^2 = 0.$

From $x(t) = e^{-t}$, $x' = -e^{-t}$, $x'' = e^{-t}$, so

$xx'' - 2(x')^2 + x^2 = e^{-t}e^{-t} - 2(-e^{-t})^2 + (e^{-t})^2 = 0.$

From $x(t) = C_1e^t + C_2e^{-t}$, $x' = C_1e^t - C_2e^{-t}$, $x'' = C_1e^t + C_2e^{-t}$, so

$xx''-2(x')^2+x^2 = (C_1e^t + C_2e^{-t})(C_1e^t + C_2e^{-t}) - 2(C_1e^t - C_2e^{-t})^2 + (C_1e^t + C_2e^{-t})^2$

$= 8C_1C_2$, which is not always zero.

With $x = 1/y$ we have $x' = -y'/y^2$ and $x'' = -y''/y^2 + 2(y')^2/y^3$, so that

$xx'' - 2(x')^2 + x^2 = 0$ becomes

$y'' - y = 0$. This has solution

$y(t) = C_1e^t + C_2e^{-t}$, so

$x(t) = 1/y(t) = 1/(C_1e^t + C_2e^{-t})$.

9. From $x(t) = 1/t$, $x' = -1/t^2$, $x'' = 2/t^3$, so

$x'' + 6xx' + 4x^3 = 2/t^3 + 6(1/t)(-1/t^2) + 4(1/t)^3 = 0.$

From $x(t) = t/(1+t^2)$, $x' = (1-t^2)/(1+t^2)^2$, $x'' = (-6t + 2t^3)/(1+t^2)^3$, so

$x'' + 6xx' + 4x^3 = (-6t+2t^3)/(1+t^2)^3 + 6t(1-t^2)/(1+t^2)^3 + 4t^3/(1+t^2)^3 = 0.$

From $x(t) = C_1/t + C_2t/(1+t^2)$,

$x' = -C_1/t^2 + C_2(1-t^2)/(1+t^2)^2$,

$x'' = 2C_1/t^3 + C_2(-6t+2t^3)/(1+t^2)^3$, and $x'' + 6xx' + 4x^3 \neq 0$.

With $x = y'/(2y)$,

$x' = y''/(2y) - (y')^2/(2y^2)$, and

$x'' = y'''/(2y) - 3y'y''/(2y^2) + 2(y')^3/(2y^3)$ then

$x'' + 6xx' + 4x^3 = y'''/(2y)$, so

$y''' = 0$. This gives

$y(t) = C_1 + C_2t + C_3t^2$, so

$x(t) = y'/(2y) = (C_2 + 2C_3t)/(2C_1 + 2C_2t + 2C_3t^2)$.

11. Figures 10-19, 10-20, 10-21, and 10-22 show the results for $x_0 = -1.0, -0.9, -0.8$, and -0.7, respectively, with periods 3, 2, 1, and 2. This exhibits sensitivity to initial conditions.

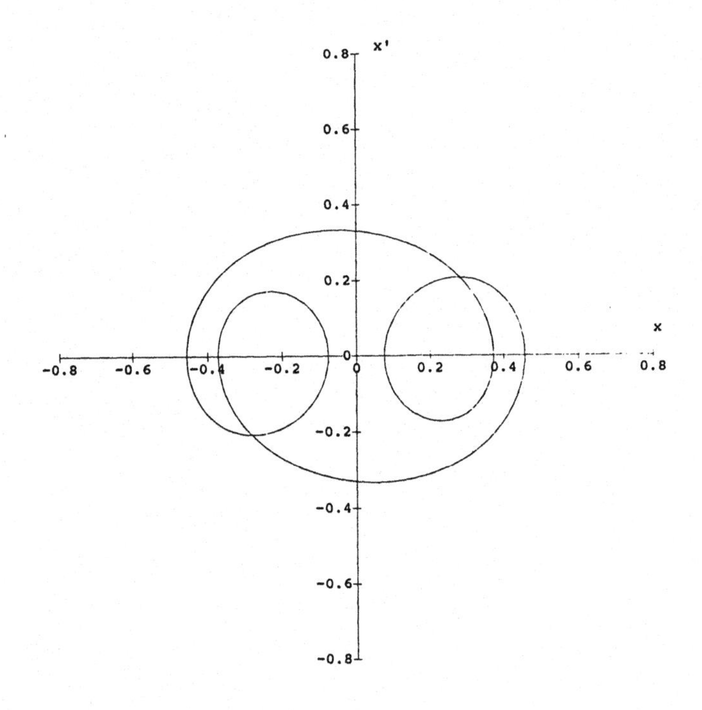

Figure 10-19 Exercise 11. $x(0) = -1$.

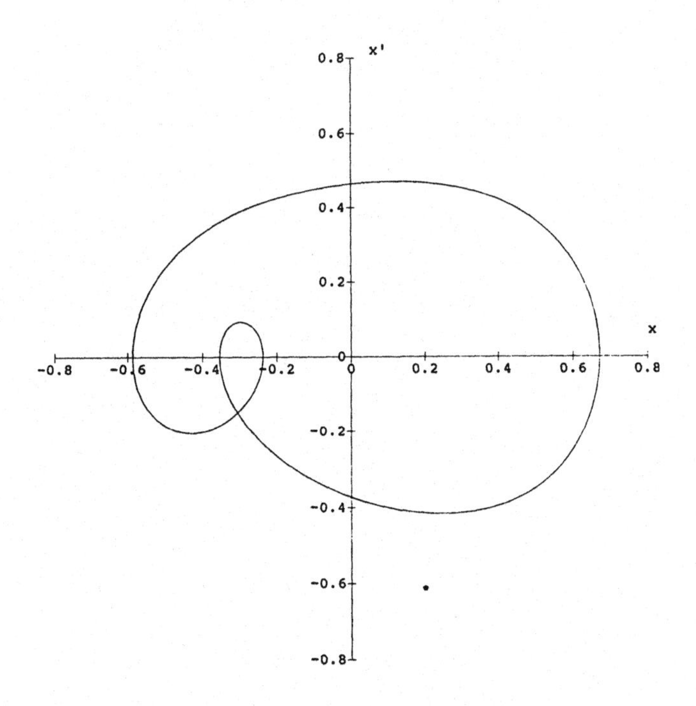

Figure 10-20 Exercise 11. $x(0) = -0.9$.

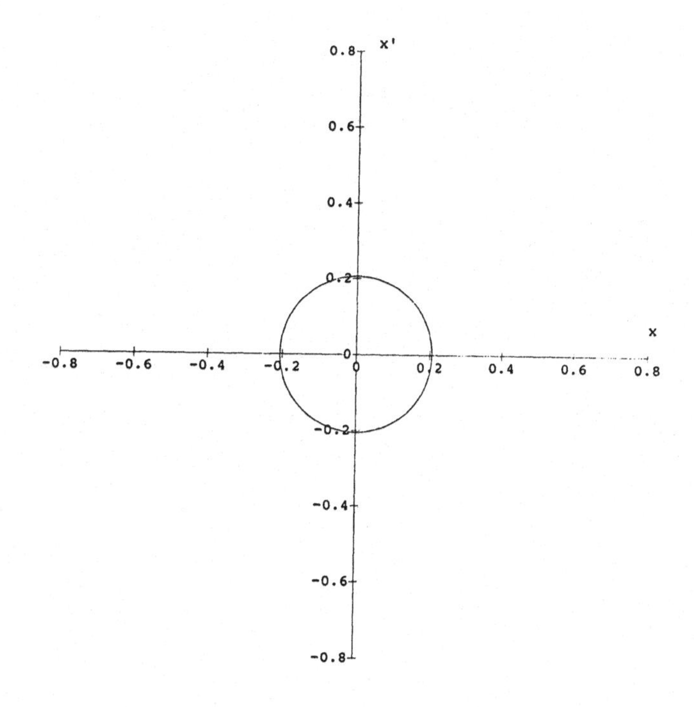

Figure 10-21 Exercise 11. $x(0) = -0.8$.

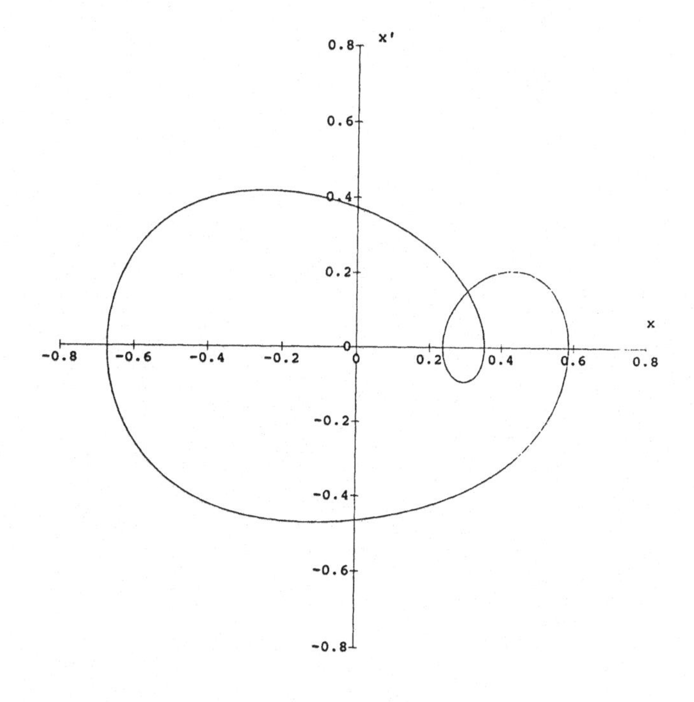

Figure 10-22 Exercise 11. $x(0) = -0.7$.

10.7 Autonomous Versus Nonautonomous Differential Equations

1. Because the differential equation is autonomous, any horizontal translation of the given solution gives another solution.

11. USING LAPLACE TRANSFORMS

11.1 Motivation

1. $\mathcal{L}\{\sin t\} = \int_0^\infty \exp(-st) \sin t \, dt$

$= \lim_{b\to\infty} \left[-\exp(-st)\cos t|_0^b\right] - s\int_0^\infty \exp(-st)\cos t \, dt$

$= 1 - s\left[\lim_{b\to\infty} exp(-st)\sin t|_0^b + s\int_0^\infty \exp(-st)\sin t \, dt\right]$

$= 1 - s^2\mathcal{L}\{\sin t\}$, so $\mathcal{L}\{\sin t\} = 1/(1+s^2)$.

3. (a) In $\mathcal{L}\{y'\} = -y(0) + s\mathcal{L}\{y\}$, let $y' = \sin at$, so $y = -(1/a)\cos at$ and $y(0) = -1/a$.
This gives $\mathcal{L}\{\sin at\} = 1/a - (s/a)\,\mathcal{L}\{\cos at\}$.
Now let $y' = \cos at$, so $y = (1/a)\sin at$ and $y(0) = 0$ and again use
$\mathcal{L}\{y'\} = -y(0) + s\mathcal{L}\{y\}$ to find
$\mathcal{L}\{\cos at\} = 0 + s\mathcal{L}\{(1/a)\sin at\}$.
Combining these results gives
$\mathcal{L}\{\sin at\} = 1/a - (s/a)\,\mathcal{L}\{\cos at\}$
$= 1/a - (s/a)\,[s\mathcal{L}\{(1/a)\sin at\}]$.
Rearranging gives $\mathcal{L}\{\sin at\} = a/(s^2 + a^2)$.

(c) In $\mathcal{L}\{y'\} = -y(0) + s\mathcal{L}\{y\}$, let $y' = \sinh at$, so $y = (1/a)\cosh at$ and $y(0) = 1/a$.
This gives $\mathcal{L}\{\sinh at\} = -1/a + (s/a)\,\mathcal{L}\{\cosh at\}$.
Now let $y' = \cosh at$, so $y = (1/a)\sinh at$ and $y(0) = 0$ and again use
$\mathcal{L}\{y'\} = -y(0) + s\mathcal{L}\{y\}$ to find
$\mathcal{L}\{\cosh at\} = [0 + s\mathcal{L}\{(1/a)\sinh at\}]$.
Combining these results gives
$\mathcal{L}\{\sinh at\} = -1/a + (s/a)\,\mathcal{L}\{\cosh at\}$
$= -1/a + (s/a)\,[s\mathcal{L}\{(1/a)\sinh at\}]$.
Rearranging gives $\mathcal{L}\{\sinh at\} = a/(s^2 - a^2)$.

5. (a) From Table 11.1 $\mathcal{L}\{\exp(2t)\} = 1/(s-2)$, $\mathcal{L}\{\exp(-3t)\} = 1/(s+3)$ so
$\mathcal{L}\{\exp(2t) + \exp(-3t)\} = 1/(s-2) + 1/(s+3)$.

(c) $\mathcal{L}\{f(t)\} = 1/[(s+2)(s+3)]$
$= A/(s+2) + B/(s+3)$
$= [A(s+3) + B(s+2)]/[(s+2)(s+3)]$.
This gives $1 = 3A + 2B$ and $0 = A + B$, or $A = 1$, $B = -1$ and
$1/[(s+2)(s+3)] = 1/(s+2) - 1/(s+3)$.
From Table 11.1 $\mathcal{L}\{\exp(-2t)\} = 1/(s+2)$, $\mathcal{L}\{\exp(-3t)\} = 1/(s+3)$, so
$f(t) = \exp(-2t) - \exp(-3t)$.

(e) $\mathcal{L}\{f(t)\} = 6/[(s-2)(s^2-1)]$
$= A/(s-2) + (Bs + C)/(s^2 - 1)$
$= [A(s^2-1) + (s-2)(Bs+C)]/[(s-2)(s^2-1)]$.
This gives $0 = A + B$, $0 = -2B + C$, $6 = -A - 2C$, so $A = 2$, $B = -2$, $C = -4$,
and $6/[(s-2)(s^2-1)] = 2/(s-2) - 2s/(s^2-1) - 4/(s^2-1)$.

From Table 11.1 $\mathcal{L}\{\exp(2t)\} = 1/(s-2)$, $\mathcal{L}\{\sinh t\} = 1/(s^2-1)$, and $\mathcal{L}\{\cosh t\} = s/(s^2-1)$, so

$f(t) = 2\exp(2t) - 2\sinh t - 4\cosh t$.

7. (a) $\mathcal{L}\{y'\} + 3\mathcal{L}\{y\} = 5\mathcal{L}\{\exp 2t\}$, so $s\mathcal{L}\{y\} + 3\mathcal{L}\{y\} = 5/(s-2)$.

$\mathcal{L}\{y\} = 5/[(s+3)(s-2)] = 1/(s-2) - 1/(s+3)$.

Thus, $y(t) = \exp 2t - \exp(-3t)$.

(c) $\mathcal{L}\{y'\} + 2\mathcal{L}\{y\} = 5\mathcal{L}\{\sin t\}$, so $s\mathcal{L}\{y\} + 2\mathcal{L}\{y\} = 5/(s^2+1)$.

$\mathcal{L}\{y\} = 5/[(s+2)(s^2+1)] = 1/(s+2) - s/(s^2+1) + 2/(s^2+1)$.

Thus, $y(t) = \exp(-2t) - \cos t + 2\sin t$.

(e) $\mathcal{L}\{y'\} + 3\mathcal{L}\{y\} = \mathcal{L}\{\exp(2t)\}$, so

$s\mathcal{L}\{y\} - y(0) + 3\mathcal{L}\{y\} = 1/(s-2)$, or

$\mathcal{L}\{y\} = 1/[(s+3)(s-2)] + y(0)/(s+3)$

$= [1/5][1/(s-2) - 1/(s+3)] + y(0)/(s+3)$.

This gives

$y(t) = [1/5][\exp(2t) - \exp(-3t)] + y(0)\exp(-3t)$.

$y(2) = 0 = [1/5][\exp(4) - \exp(-6)] + y(0)\exp(-6)$, so $y(0) = [1-\exp(10)]/5$.

Thus,

$y(t) = [\exp(2t) - \exp(-3t)]/5 + [1-\exp(10)][\exp(-3t)]/5$.

11.2 Constructing New Laplace Transforms from Old

1. (a) $f(t) = t\cos at$,

$f'(t) = \cos at - at\sin at$,

$f''(t) = -2a\sin at - a^2 t\cos at$.

$f(0) = 0$, $f'(0) = 1$.

$\mathcal{L}\{f''(t)\} = s^2\mathcal{L}\{f(t)\} - sf(0) - f'(0)$ so $\mathcal{L}\{-2a\sin at - a^2 t\cos at\} = s^2\mathcal{L}\{t\cos at\} - 1$.

Rearranging gives

$(s^2+a^2)\mathcal{L}\{t\cos at\} = 1 - 2a\mathcal{L}\{\sin at\} = 1 - 2a^2/(s^2+a^2)$, so

$\mathcal{L}\{t\cos at\} = 1/(s^2+a^2) - 2a^2/(s^2+a^2)^2$

$= (s^2-a^2)/(s^2+a^2)^2$.

3. (a) $f(t) = t\sinh t$,

$f'(t) = \sinh t + t\cosh t$,

$f''(t) = 2\cosh t + t\sinh t$.

$f(0) = 0$, $f'(0) = 0$.

$\mathcal{L}\{f''(t)\} = s^2\mathcal{L}\{f(t)\} - sf(0) - f(0)$ so $\mathcal{L}\{2\cosh t + t\sinh t\} = s^2\mathcal{L}\{t\sinh t\}$.

Rearranging gives

$(s^2-1)\mathcal{L}\{t\sinh t\} = 2\mathcal{L}\{\cosh t\} = 2s/(s^2-1)$, so

$\mathcal{L}\{t\sinh t\} = 2s/(s^2-1)^2$.

5. (a) If $f(t) = \sinh at$, $\mathcal{L}\{f(t)\} = a/(s^2-a^2)$, so by the Shifting Theorem,

$\mathcal{L}\{\exp(at)f(t)\} = a/[(s-a)^2 - a^2]$

$= a/[s^2 - 2sa]$.

(c) $\mathcal{L}\{t^2\} = 2/s^3$, so by the Shifting Theorem,
$\mathcal{L}\{\exp(at)t^2\} = 2/(s-a)^3$.

7. (a) $\mathcal{L}\{\cos at\} = s/(s^2+a^2) = F(s)$, so by the Special Product Theorem,
$\mathcal{L}\{t^2 \cos at\} = F''(s)$
$= (-1)^2 d/ds[1/(s^2+a^2) - 2s^2/(s^2+a^2)^2]$
$= -2s/(s^2+a^2)^2 - 4s/(s^2+a^2)^2 - [2s^2][-4s/(s^2+a^2)^3]$
$= -6s/(s^2+a^2)^2 + 8s^3/(s^2+a^2)^3$
$= (2s^3 - 6a^2 s)/(s^2+a^2)^3$

(c) $\mathcal{L}\{\cosh at\} = s/(s^2-a^2) = F(s)$, so by the Special Product Theorem,
$\mathcal{L}\{t \cosh at\} = (-1)d/ds[s/(s^2-a^2)]$
$= -1/(s^2-a^2) + 2s^2/(s^2-a^2)^2$
$= (s^2+a^2)/(s^2-a^2)^2$.

11.3 The Inverse Laplace Transform and the Convolution Theorem

1. (a) $1/(s^2-3s+2)$
$= 1/[(s-3/2)^2 - 9/4 + 2]$
$= 1/[(s-3/2)^2 - 1/4]$ and $\mathcal{L}^{-1}\{1/(s^2-1/4)\} = 2\sinh t/2$, so by the Shifting Theorem, $\mathcal{L}^{-1}\{1/[(s-3/2)^2 - 1/4)]\} = 2\exp(3t/2)\sinh t/2 = \exp 2t - \exp t$.
Alternatively,
$1/(s^2-3s+2) = 1/(s-2) - 1/(s-1)$, so $\mathcal{L}^{-1}\left\{1/(s^2-3s+2)\right\} = \exp 2t - \exp t$.

(c) $1/(s^3+2s^2-s-2)$
$= 1/[s^2(s+2) - (s+2)]$
$= 1/[(s^2-1)(s+2)]$
$= (As+B)/(s^2-1) + C/(s+2)$
$= [(As+B)(s+2) + C(s^2-1)]/[(s^2-1)(s+2)]$.
This gives $0 = A + C$, $0 = 2A + B$, and $1 = 2B - C$, so $A = -1/3$, $B = 2/3$, $C = 1/3$, so
$1/(s^3+2s^2-s-2) = (-s/3+2/3)/(s^2-1) + (1/3)/(s+2)$, and
$\mathcal{L}^{-1}\{1/(s^3+2s^2-s-2)\} = -(1/3)\cosh t + 2/3 \sinh t + (1/3)\exp(-2t)$.

(e) $s/[(s-1)(s^2+a^2)$
$= A/(s-1) + (Bs+C)/(s^2+a^2)$
$= [A(s^2+a^2) + (Bs+C)(s-1)]/[(s-1)(s^2+a^2)]$.
This gives $0 = A + B$, $1 = C - B$, and $0 = Aa^2 - C$, so $A = 1/(a^2+1)$, $B = -1/(a^2+1)$, and $C = a^2/(a^2+1)$ and
$\mathcal{L}^{-1}\{s/[(s-1)(s^2+a^2)\} = 1/(a^2+1)\exp t - 1/(a^2+1)\cos at + a/(a^2+1)\sin at$.

(g) $1/[s(s^2+4s+13)]$
$= 1/[s(s^2+4s+4+9)]$
$= 1/[s\{(s+2)^2+9\}]$.
$\mathcal{L}\{1\} = 1/s$, and $\mathcal{L}\{\exp(-2t)(1/3)\sin 3t\} = 1/[(s+2)^2+9]$, so by the Convolution Theorem we have
$\mathcal{L}^{-1}\{1/[s\{(s+2)^2+9\}]\}$
$= \int_0^t \exp(-2z)(1/3)\sin 3z(1)\,dz$

$= \left\{ (1/3)\exp(-2z)[-2\sin 3z - 3\cos 3z]/[4+9]|_0^t \right\}$
$= (1/3)(\exp(-2t)[-2\sin 3t - 3\cos 3t] + 3)/[13]$
$= (1/39)(\exp(-2t)[-2\sin 3t - 3\cos 3t] + 3).$

(i) From Table 11.6 $\mathcal{L}^{-1}\{a/[(s-1)(s^2+a^2)] = (a\exp t - \sin at - a\cos at)/(1+a^2).$

(k) This is listed in Table 11.6.

(m) This is listed in Table 11.6.

3. From (11.45), if $f(t) = t\exp t - \exp t + 1$, then $\mathcal{L}\{f(t)\} = 1/[s(s-1)^2]$.
From (11.38) $\mathcal{L}^{-1}\{F(s/a)\} = af(at)$, so
$\mathcal{L}^{-1}\{1/[(s/a)(s/a-1)^2]\} = a[at\exp at - \exp at + 1]$ or, dividing both sides by a^3 gives
$\mathcal{L}^{-1}\{1/[(s)(s-a)^2]\} = (1/a^2)[at\exp at - \exp at + 1]$.

5. **(a)** $\mathcal{L}\{y\} = \mathcal{L}\{4t-3\} - \mathcal{L}\{\int_0^t \sin(t-z)y(z)\,dz\}$
$= 4/s^2 - 3/s - \mathcal{L}\{\sin t\}\mathcal{L}\{y\}$
$= 4/s^2 - 3/s - [1/(s^2+1)][\mathcal{L}\{y\}]$.
Solving for $\mathcal{L}\{y\}$ gives
$\mathcal{L}\{y\} = [4/s^2 - 3/s](s^2+1)/(s^2+2)$
$= (4-3s)(s^2+1)/[s^2(s^2+2)]$.
Using partial fractions gives
$(4-3s)(s^2+1)/[s^2(s^2+2)]$
$= (A+Bs)/s^2 + (Cs+D)/(s^2+2)$
$= [(A+Bs)(s^2+2) + (Cs+D)s^2]/[s^2(s^2+2)]$.
Equating coefficients of like powers of s gives $4 = 2A$, $-3 = 2B$, $4 = A+D$, and $-3 = B+C$. Solving gives $A = 2$, $B = -3/2$, $C = -3/2$, and $D = 2$, so
$\mathcal{L}\{y\} = [2/s^2 - (3/2)/s + (-3s/2+2)/(s^2+2)]$ and
$y(t) = 2t - 3/2 - (3/2)\cos(\sqrt{2}t) + \sqrt{2}\sin(\sqrt{2}t)$.

(c) $\mathcal{L}\{y\} = 3\mathcal{L}\{\sin t\} - \mathcal{L}\{2\int_0^t \cos(t-z)y(z)\,dz\}$
$= 3/\left(s^2+1\right) - 2\mathcal{L}\{\cos t\}\mathcal{L}\{y\}$
$= 3/\left(s^2+1\right) - 2\left[s/\left(s^2+1\right)\right]\mathcal{L}\{y\}$
Solving for $\mathcal{L}\{y\}$ gives
$\mathcal{L}\{y\} = 3/(s+1)^2$ and
$y(t) = 3t\exp(-t)$.

(e) $\mathcal{L}\{y\} = \mathcal{L}\{t\} - \mathcal{L}\{\int_0^t \exp(t-z)y(z)\,dz\}$
$= 1/s^2 - \mathcal{L}\{\exp(t)\}\mathcal{L}\{y\}$
$= 1/s^2 - [1/(s-1)]\mathcal{L}\{y\}$.
Solving for $\mathcal{L}\{y\}$ gives
$\mathcal{L}\{y\} = 1/\left\{s^2[1 + 1/(s-1)]\right\} = (s-1)/s^3 = 1/s^2 - 1/s^3$ and
$y(t) = t - t^2/2$.

11.4 Functions That Jump

1. **(a)** $f(t) = u(t) - u(t-4)$.

(c) $f(t) = t[u(t) - u(t-2)] + 2[u(t-2) - u(t-4)] + (6-t)[u(t-4) - u(t-6)]$.

(e) $f(t) = [u(t) - u(t-a)] - [u(t-2a) - u(t-3a)]$, where $f(t+3a) = f(t)$.

3. $g(t) = \sin t[u(t) - u(t-\pi)]$, $g(t+\pi) = g(t)$.

$\mathcal{L}\{g(t)\} = [1/(1-\exp(-\pi s))] \int_0^\pi \exp(-st) \sin t\, dt$

$= [1/(1-\exp(-\pi s))]\{(\exp(-st)/[1+s^2])\, (-s \sin t - \cos t)|_0^\pi\}$

$= [1/(1-\exp(-\pi s))]\, [\exp(-s\pi) + 1]\, /[1+s^2])$.

5. (a) $\mathcal{L}\{f(t)\} = \int_0^2 \exp(-st)\, t dt$

$= \{\exp(-st)(-t/s) - \exp(-st)/s^2 |_0^2\}$

$= \exp(-2s)(-2/s) - \exp(-2s)/s^2 + 1/s^2$

$= [1 - \exp(-2s)]/s^2 - 2\exp(-2s)/s$.

Alternatively,

$\mathcal{L}\{f(t)\} = \mathcal{L}\{t\, [u(t) - u(t-2)]\}$

$= \mathcal{L}\{tu(t)\} - \mathcal{L}\{tu(t-2)\}$

$= 1/s^2 - \mathcal{L}\{(t-2+2)\, u(t-2)\}$

$= 1/s^2 - \mathcal{L}\{(t-2)\, u(t-2)\} - \mathcal{L}\{2u(t-2)\}$

$= 1/s^2 - \exp(-2s)/s^2 - 2\exp(-2s)/s$.

(c) $\mathcal{L}\{f(t)\} = [1/(1-\exp(-2s))] \int_0^2 \exp(-st)t\, dt$

$= [1/(1-\exp(-2s))]\{[1-\exp(-2s)]/s^2 - 2\exp(-2s)/s\}$

$= 1/s^2 - (2/s)/(\exp 2s - 1)$.

7. (a) $\mathcal{L}^{-1}\{(s+1)^2\} = t\exp(-t)$, so

$\mathcal{L}^{-1}\{\exp(-s)(s+1)^2\} = (t-1)\exp(-\{t-1\})u(t-1)$.

(c) $\mathcal{L}^{-1}\{1/[s(s+1)]\} = 1 - \exp(-t)$, so

$\mathcal{L}^{-1}\{\exp(-3s)/[s(s+1)]\} = [1 - \exp(-\{t-3\})]u(t-3)$.

(e) $\mathcal{L}^{-1}\{1/(s-4)^2\} = t\exp 4t$, so

$\mathcal{L}^{-1}\{\exp(-s)/(s-4)^2 - \exp(-2s)/(s-4)^2\}$

$= (t-1)\exp(4[t-1])u(t-1) - (t-2)\exp(4[t-2])u(t-2)$.

(g) $1/(s^2+6s+25) = 1/[(s+3)^2+16\}$, so

$\mathcal{L}^{-1}\{1/(s^2+6s+25)\} = (1/4)\exp(-3t)\sin 4t$.

Using (11.47) gives

$\mathcal{L}^{-1}\{\exp(-3s)/(s^2+6s+25)\} = (1/4)\exp(-3[t-3])\sin[4(t-3)]\, u(t-3)$.

11.5 Models Involving First Order Linear Differential Equations

1. (a) $\mathcal{L}\{(\exp(kt)-1)/k\} = 1/[s(s-k)]$, and $\mathcal{L}^{-1}\{(1/s)F(s)\} = \int_0^t f(z)dz$, so

$\mathcal{L}^{-1}\{(1/s)(1/[s(s-k)]\} = \int_0^t (\exp(kz) - 1)/k\, dz$

$= \{\exp(kz)/k^2 - z/k\}|_0^t$

$= (\exp kt - 1)/k^2 - t/k$.

3. (a) $\mathcal{L}\{y' + y\} = \mathcal{L}\{2u(t) - u(t-1)\}$, giving $sY(s) + Y(s) = 2/s - \exp(-s)/s$, so $Y(s) = 2/[s(s+1)] - \exp(-s)/[s(s+1)]$. Finding inverse Laplace transforms gives $y(t) = 2[1-\exp(-t)] - [1 - \exp\{-(t-1)\}]u(t-1)$.

(c) $\mathcal{L}\{y'+y\} = \mathcal{L}\{5u(t) - 4u(t-10)\}$, giving
$sY(s) - 6 + Y(s) = 5/s - 4\exp(-10s)/s$, so
$Y(s) = 6/(s+1) + 5/[s(s+1)] - 4\exp(-10s)/[s(s+1)]$.
Finding inverse Laplace transforms gives
$y(t) = 6\exp(-t) + 5[1-\exp(-t)] - 4[1-\exp(-\{t-10\})]u(t-10)$.

5. (a) $\mathcal{L}\{LI' + RI\} = E_0[\mathcal{L}\{u(t-1)\} - \mathcal{L}\{u(t-2)\}]$, or
$sL\mathcal{L}\{I\} + R\mathcal{L}\{I\} = E_0[\exp(-s)/s - \exp(-2s)/s]$.
Solving for $\mathcal{L}\{I\}$ gives
$\mathcal{L}\{I\} = E_0[\exp(-s)/s - \exp(-2s)/s]/[sL+R]$.
Now $\mathcal{L}^{-1}\{1/[s(sL+R)]\} = [1-\exp(-Rt/L)]/R$, so
$I(t) = [1-\exp(-R[t-1]/L)]u(t-1)/R - [1-\exp(-R[t-2]/L)]u(t-2)/R$.

(c) $sL\mathcal{L}\{I\} + R\mathcal{L}\{I\} = E_0\mathcal{L}\{f(t)\} = 1/s\sum_{k=0}^{\infty}(-1)^k\exp(-sk)$.
Solving for $\mathcal{L}\{I\}$ gives
$\mathcal{L}\{I\} = E_0[\sum_{k=0}^{\infty}(-1)^k\exp(-sk)]/[s(sL+R)]$.
Now $\mathcal{L}^{-1}\{1/[s(sL+R)]\} = [1-\exp(-Rt/L)]/R$, so
$\mathcal{L}^{-1}\{\exp(-sk)/[s(sL+R)\} = [1-\exp(-R[t-k]/L)]u(t-k)/R$ and
$I(t) = \sum_{k=0}^{\infty}\{(-1)^k[1-\exp(-R[t-k]/L)]u(t-k)/R\}$.

7. (a) $T_a(t) = 425[u(t) - u(t-15)] + [425 - 7.5(t-15)][u(t-15) - u(t-25)] + 350u(t-25$
$= 425u(t) - 7.5(t-15)u(t-15) + 7.5(t-25)u(t-25)$

(b) $T' = k[T - T_a(t)]$ gives
$s\mathcal{L}\{T\} - T(0) = k\left[\mathcal{L}\{T\} - 425/s + 7.5\exp(-15s)/s^2 - 7.5\exp(-25s)/s^2\right]$.
Solving for $\mathcal{L}\{T\}$ gives
$\mathcal{L}\{T\} = \left\{70 + k\left[-425/s + 7.5\exp(-15s)/s^2 - 7.5\exp(-25s)/s^2\right]\right\}/(s-k)$.
Now $\mathcal{L}^{-1}\{k/[s(s-k)]\} = e^{kt} - 1$, so
$\mathcal{L}^{-1}\{1/[s^2(s-k)]\} = \int_0^t (e^{kz} - 1)\,dz = (e^{kt} - 1 - kt)/k$.
$T(t) = 425 - 355e^{kt} + 7.5[f(t-15)u(t-15) - f(t-25)u(t-25)]$,
where $f(t) = (e^{kt} - 1 - kt)/k$.

(c) $T_a(t) = 425[u(t) - u(t-15)] + 350u(t-15)$
$= 425u(t) - 75u(t-15)$.
$T' = k[T - T_a(t)]$ gives
$s\mathcal{L}\{T\} - T(0) = k[\mathcal{L}\{T\} - 425/s + 75\exp(-15s)/s]$.
Solving for $\mathcal{L}\{T\}$ gives
$\mathcal{L}\{T\} = \{70 - k[425/s - 75\exp(-15s)/s]\}/(s-k)$.
$T(t) = 70 - 425(e^{kt} - 1)u(t) + 75\left[e^{k(t-15)} - 1\right]u(t-15)$
$= 425 - 355e^{kt} + 75\left[e^{k(t-15)} - 1\right]u(t-15)$.

(d) The solutions are identical for $0 < t < 15$. For $t > 15$, the solution from part (b) exceeds 350° and then is asymptotic to $T(t) = 350$. The solution from part (c) approaches 350° from below.

11.6 Models Involving Higher Order Linear Differential Equations

1. (a) $s^2Y(s) - sy(0) - y'(0) + sY(s) - y(0) - 2Y(s) = 0$, or solving for $Y(s)$,

$Y(s) = 3/[s^2 + s - 2]$
$= 1/(s-1) - 1/(s+2)$.
Taking inverse Laplace transforms gives
$y(t) = \exp t - \exp(-2t)$.

(c) $s^2Y(s) - sy(0) - y'(0) + 6sY(s) - 6y(0) + 9Y(s) = 0$, or solving for $Y(s)$,
$Y(s) = [2s + 12]/[s^2 + 6s + 9]$.
Because we know the inverse Laplace transforms of both $1/(s+3)$ and $1/(s+3)^2$, but not that of $s/(s+3)^2$, we rewrite $[2s+12]/[s^2+6s+9]$ as
$[2s + 6 + 6]/(s+3)^2 = 2/(s+3) + 6/(s+3)^2$.
Taking inverse Laplace transforms gives
$y(t) = 2\exp(-3t) + 6t\exp(-3t)$.

(e) $s^2Y(s) - sy(0) - y'(0) + 3sY(s) - 3y(0) = 0$, or solving for $Y(s)$,
$Y(s) = [4s + 15]/[s^2 + 3s]$
$= 4/(s+3) + 15/[s(s+3)]$.
Taking inverse Laplace transforms gives
$y(t) = 4\exp(-3t) + 15\,[1 - \exp(-3t)]\,/3 = 5 - \exp(-3t)$.

(g) $s^2Y(s) - sy(0) - y'(0) + 10sY(s) - 10y(0) + 100Y(s) = 0$, or solving for $Y(s)$,
$Y(s) = [15s + 154]/[s^2 + 10s + 100]$
$= 15\,(s+5)\,/\,\left[(s+5)^2 + 75\right] + 79/\left[(s+5)^2 + 75\right]$.
Taking inverse Laplace transforms gives
$y(t) = 15e^{-5t}\cos\left(5\sqrt{3}t\right) + \left(79\sqrt{3}/15\right)e^{-5t}\sin\left(5\sqrt{3}t\right)$.

3. **(a)** $s^2Y(s) - sy(0) - y'(0) + sY(s) - y(0) - 12Y(s) = 8/(s-3)$, or solving for $Y(s)$,
$Y(s) = [1 + 8/(s-3)]/[s^2 + s - 12]$
$= 1/[(s+4)(s-3)] + 8/[(s+4)(s-3)^2]$.
Now
$\mathcal{L}^{-1}\{1/[(s+4)(s-3)]\} = (1/7)\,[\exp 3t - \exp(-4t)]$ and
$\mathcal{L}^{-1}\{1/(s-3)\} = \exp 3t$, so by the Convolution Theorem
$\mathcal{L}^{-1}\{8/[(s+4)(s-3)^2]\}$
$= (8/7)\int_0^t \exp(3(t-z)[\exp(3z) - \exp(-4z)]\,dz$
$= (8/7)\,[t\exp 3t + \exp 3t\{\exp(-7t) - 1\}/7]$.
Combining these two results gives
$y(t) = (8t/7)\exp 3t + (1/49)\,[\exp(-4t) - \exp(3t)]$.

(c) $s^2Y(s) - 5sY(s) + 6Y(s) = 12/(s+1)^2 - 7/(s+1)$, or solving for $Y(s)$,
$Y(s) = [12/(s+1)^2 - 7/(s+1)]/[s^2 - 5s + 6]$
$= 12/[(s-2)(s-3)(s+1)^2] - 7/[(s-2)(s-3)(s+1)]$.
Now
$7/[(s-2)(s-3)(s+1)]$
$= A/(s-2) + B/(s-3) + C/(s+1)$
$= [A(s-3)(s+1) + B(s-2)(s+1) + C(s-2)(s-3)]/[(s-2)(s-3)(s+1)]$.
This requires that $7 = A(s-3)(s+1) + B(s-2)(s+1) + C(s-2)(s-3)]$ be valid for all values of s.
Setting $s = 3, -1$, and 2, gives (in order) $7 = 4B$, $7 = 12C$, and $7 = -3A$, so $A = -7/3$, $B = 7/4$, and $C = 7/12$.
Thus,
$\mathcal{L}^{-1}\{7/[(s-2)(s-3)(s+1)]\} = \{7\}\{-\exp(2t)/3 + \exp(3t)/4 + \exp(-t)/12\}$,
and by the Convolution Theorem

$\mathcal{L}^{-1}\{12/[(s-2)(s-3)(s+1)^2]\}$
$= 12\int_0^t \exp(-(t-z))[-\exp(2z)/3 + \exp(3z)/4 + \exp(-z)/12]\,dz$
$= 12e^{-t}\int_0^t[-\exp(3z)/3 + \exp(4z)/4 + 1/12]\,dz$
$= t\exp(-t) - (4/3)\exp 2t + (3/4)\exp 3t + (7/12)\exp(-t).$
Combining these two results gives
$y(t) = t\exp(-t) + \exp(2t) - \exp(3t).$

5. **(a)** $s^2\mathcal{L}\{x\} + \mathcal{L}\{x\} = \mathcal{L}\{\exp(-2t)\sin t\}$, or solving for $\mathcal{L}\{x\}$,
$\mathcal{L}\{x\} = \mathcal{L}\{\exp(-2t)\sin t\}/(s^2+1)\}.$
Using the convolution theorem, we find
$x(t) = \int_0^t \exp(-2z)\sin z\sin(t-z)\,dz,$
or using $2\sin a\sin b = \cos(a-b) - \cos(a+b)$ gives
$x(t) = \int_0^t \exp(-2z)[\cos(2z-t) - \cos t]\,dz.$
Integration gives $x(t) = \exp(-2t)[\cos t + \sin t]/8 + [\sin t - \cos t]/8.$

(c) $s^2\mathcal{L}\{x\} + 4s\mathcal{L}\{x\} + 4\mathcal{L}\{x\} = 1/(s+3)$, or solving for $\mathcal{L}\{x\}$,
$\mathcal{L}\{x\} = 1/[(s+3)(s+2)^2]$
$= A/(s+3) + B/(s+2) + C/(s+2)^2$
$= [A(s+2)^2 + B(s+2)(s+3) + C(s+3)]/[(s+3)(s+2)^2].$
Equating coefficients of like powers of s gives $A+B=0$, $4A+5B+C=0$, and $4A+6B+3C=1$, so $A=1$, $B=-1$, and $C=1$, and
$x(t) = \exp(-3t) - \exp(-2t) + t\exp(-2t).$

(e) $s^2\mathcal{L}\{x\} + 3s\mathcal{L}\{x\} + 2\mathcal{L}\{x\} = \mathcal{L}\{t^2[1-u(t-1)]\}$, or solving for $\mathcal{L}\{x\}$,
$\mathcal{L}\{x\} = \mathcal{L}\{t^2[1-u(t-1)]\}/[(s+2)(s+1)].$
Using the convolution theorem gives
$x(t) = \int_0^t \left[e^{-(t-z)} - e^{-2(t-z)}\right] z^2[1-u(z-1)]\,dz$
$= e^{-t}\int_0^t e^z z^2[1-u(z-1)]\,dz - e^{-2t}\int_0^t e^{2z}z^2[1-u(z-1)]\,dz.$
If $0<t<1$ then
$x(t) = e^{-t}\int_0^t e^z z^2\,dz - e^{-2t}\int_0^t e^{2z}z^2\,dz$
$= t^2 - 2t + 2 - 2e^{-t} - \left(2t^2 - 2t + 1 - e^{-2t}\right)/4.$
If $1<t$ then
$x(t) = e^{-t}\int_0^1 e^z z^2\,dz - e^{-2t}\int_0^1 e^{2z}z^2\,dz$
$= e^{-t}(e-2) - e^{-2t}\left(e^2-1\right)/4.$

(g) $s^2\mathcal{L}\{x\} - s + 4s\mathcal{L}\{x\} - 4 + 4\mathcal{L}\{x\} = \mathcal{L}\{f(t)\}$, or solving for $\mathcal{L}\{x\}$,
$\mathcal{L}\{x\} = \mathcal{L}\{f(t)\}/(s+2)^2 + (s+2+2)/(s+2)^2$
$= \mathcal{L}\{f(t)\}/(s+2)^2 + 1/(s+2) + 2/(s+2)^2.$
Using the convolution theorem gives
$x(t) = \int_0^t z\exp(-2z)f(t-z)\,dz + \exp(-2t) + 2t\exp(-2t).$

(i) $s^2\mathcal{L}\{x\} - 1 + \omega^2\mathcal{L}\{x\} = \mathcal{L}\{f(t)\}$, or solving for $\mathcal{L}\{x\}$,
$\mathcal{L}\{x\} = \mathcal{L}\{f(t)\}/(s^2+\omega^2) + 1/(s^2+\omega^2).$
Using the convolution theorem gives
$x(t) = \int_0^t [\sin(\omega z)/\omega]\,f(t-z)\,dz + \sin(\omega t)/\omega.$

(k) $s^2\mathcal{L}\{x\} - 1 + \omega^2\mathcal{L}\{x\} = \mathcal{L}\{u(t-1) - u(t-2)\}$, or solving for $\mathcal{L}\{x\}$,
$\mathcal{L}\{x\} = 1/(s^2+\omega^2) + \mathcal{L}\{u(t-1) - u(t-2)\}/(s^2+\omega^2).$
Using the convolution theorem gives
$x(t) = \sin(\omega t)/\omega + \int_0^t [\sin(\omega(t-z))/\omega]\,[u(t-1) - u(t-2)]\,dz$

$$= \begin{cases} \sin(\omega t)/\omega & 0<t<1 \\ \sin(\omega t)/\omega + [1 - \cos(\omega(t-1))]/\omega^2 & 1<t<2 \\ \sin(\omega t)/\omega + [\cos(\omega(t-2)) - \cos(\omega(t-1))]/\omega^2 & 2<t. \end{cases}$$

7. **(a)** $q'' + 4q' + 20q = 10\exp(-t/2)$, so

$s^2\mathcal{L}\{q\} - s + 4s\mathcal{L}\{q\} - 4 + 20\mathcal{L}\{q\} = 10/(s+1/2)$.

Solving for $\mathcal{L}\{q\}$ gives

$\mathcal{L}\{q\} = 10/[(s+1/2)(\{s+2\}^2+16)] + (s+2+2)/(\{s+2\}^2+16)$.

By the convolution theorem

$q(t) = 10\int_0^t [\exp(-2z)\sin(4z)/4]\exp(-[t-z]/2)\,dz + \exp(-2t)[\cos(4t)+\sin(4t)/2]$

$= (5/2)\exp(-t/2)\int_0^t \exp(-3z/2)\sin 4z\,dz + \exp(-2t)[\cos(4t)+\sin(4t)/2]$

$= (5/2)\exp(-t/2)\,[\exp(-3z/2)/(16+9/4)\{-(3/2)\sin 4z - 4\cos 4z\}]|_0^t +$

$+ \exp(-2t)[\cos(4t)+\sin(4t)/2]$

$= (5/2)\exp(-t/2)[\exp(-3t/2)/(16+9/4)\{-(3/2)\sin 4t - 4\cos 4t\} + 16/73] +$

$+ \exp(-2t)[\cos(4t)+\sin(4t)/2]$

$= (10/73)\exp(-2t)[-3/2\sin 4t - 4\cos 4t] + (40/73)\exp(-t/2)$.

(c) $q'' + 5q' + 6q = 2[1-u(t-2)]$, so

$s^2\mathcal{L}\{q\} - s + 5s\mathcal{L}\{q\} - 5 + 6\mathcal{L}\{q\} = 2[1-\exp(-2s)]/s$.

Solving for $\mathcal{L}\{q\}$ gives

$\mathcal{L}\{q\} = 2[1-\exp(-2s)]/[s(s+2)(s+3)] + (s+5)/[(s+2)(s+3)]$.

Now $2/[s(s+2)(s+3)]$

$= A/s + B/(s+2) + C/(s+3)$

$= [A(s+2)(s+3) + B(s)(s+3) + C(s)(s+2)]/[s(s+2)(s+3)]$.

This requires that $2 = A(s+2)(s+3) + B(s)(s+3) + C(s)(s+2)$ be valid for all values of s.

Setting $s = 0$, -2 and -3 gives (in order) $2 = 6A$, $2 = -2B$, and $2 = 3C$,or $A = 1/3$, $B = -1$, and $C = 2/3$.

This gives

$\mathcal{L}^{-1}\{2/[s(s+2)(s+3)]\} = 1/3 - \exp(-2t) + 2/3\exp(-3t)$.

Using (11.47) now gives

$\mathcal{L}^{-1}\{2[1-\exp(-2s)]/[s(s+2)(s+3)]\}$

$= f(t) - f(t-2)u(t-2)$, where $f(t) = 1/3 - \exp(-2t) + 2/3\exp(-3t)$.

Also, $(s+5)/[(s+2)(s+3)] = 3/(s+2) - 2/(s+3)$ so

$\mathcal{L}^{-1}\{(s+5)/[(s+2)(s+3)]\} = 3\exp(-2t) - 2\exp(-3t)$ and

$q(t) = 3\exp(-2t) - 2\exp(-3t) + f(t) - f(t-2)u(t-2)$.

(e) $q'' + 5q' + 6q = E(t)$, where $E(t) = 2[1-u(t-2)]$, and $E(t+4) = E(t)$, so

$s^2\mathcal{L}\{q\} - s + 5s\mathcal{L}\{q\} - 5 + 6\mathcal{L}\{q\} = 2[1-\exp(-2s)]/\{s[1-\exp(-4s)]\}$

$= 2/\{s[1+\exp(-2s)]\}$.

Solving for $\mathcal{L}\{q\}$ gives

$\mathcal{L}\{q\} = 2/\{s(s+2)(s+3)[1+\exp(-2s)]\} + (s+5)/[(s+2)(s+3)]$.

Expanding $1/[1+\exp(-2s)]$ gives

$1/[1+\exp(-2s)] = \sum_{n=0}^{\infty}(-1)^n e^{-2ns}$, while from Exercise 7(c),

$\mathcal{L}^{-1}\{2/[s(s+2)(s+3)]\} = 1/3 - \exp(-2t) + 2/3\exp(-3t) = f(t)$, and

$\mathcal{L}^{-1}\{(s+5)/[(s+2)(s+3)]\} = 3\exp(-2t) - 2\exp(-3t)$.

Combining these results gives

$q(t) = 3\exp(-2t) - 2\exp(-3t) + \sum_{n=0}^{\infty}(-1)^n f(t-2n)u(t-2n)$.

11.7 Applications to Systems of Linear Differential Equations

1. $2/[s^2(s+4)]$

$= A/s + B/s^2 + C/(s+4)$

$= [As(s+4) + B(s+4) + Cs^2]/[s^2(s+4)].$

Equating coefficients of like powers of s gives $0 = A + C$, $0 = 4A + B$,and $2 = 4B$, with solution $B = 1/2$, $A = -1/8$, and $C = 1/8$.

Thus,

$\mathcal{L}^{-1}\{2/[s^2(s+4)]\} = -1/8 + t/2 + 1/8\exp(-4t)$, as given in Example 14.31.

3. **(a)** Taking Laplace transforms of both equations and using the result $\mathcal{L}\{y'\} = s\mathcal{L}\{y\} - y(0)$ together with the initial conditions gives

$sX(s) - 2 = 2X(s) - Y(s),$

$sY(s) = -X(s) + 2Y(s)$, or

$(s-2)X(s) + Y(s) = 2,$

$X(s) + (s-2)Y(s) = 0.$

Solving this system of algebraic equations gives

$X(s) = 2(s-2)/[(s-3)(s-1)],$

$Y(s) = -2/[(s-3)(s-1)].$

Using partial fractions gives

$Y(s) = -2/[(s-3)(s-1)]$

$= 1/(s-1) - 1/(s-3)$, so

$y(t) = \exp t - \exp 3t.$

Now write

$X(s) = 2(s-1-1)/[(s-3)(s-1)]$

$= 2/(s-3) - 2/[(s-3)(s-1)]$, so

$x(t) = 2\exp 3t + \exp t - \exp 3t = \exp 3t + \exp t.$

(c) Taking Laplace transforms of both equations and using the result $\mathcal{L}\{y'\} = s\mathcal{L}\{y\} - y(0)$ together with the initial conditions gives

$sX(s) - 2 = 3X(s) - 2Y(s),$

$sY(s) - 1 = 2X(s) - 2Y(s)$, or

$(s-3)X(s) + 2Y(s) = 2,$

$-2X(s) + (s+2)Y(s) = 1.$

Solving this system of algebraic equations gives

$X(s) = 2/(s-2),$

$Y(s) = 1/(s-2),$

so $x(t) = 2\exp 2t$, $y(t) = \exp 2t$.

5. **(a)** Taking Laplace transforms of both equations and using the result $\mathcal{L}\{y'\} = s\mathcal{L}\{y\} - y(0)$ together with the initial conditions gives

$sX(s) - 1 = -2X(s) + 4Y(s) + 7/(s-3),$

$sY(s) = X(s) + Y(s) - 2/(s-3)$, or

$(s+2)X(s) - 4Y(s) = 1 + 7/(s-3),$

$-X(s) + (s-1)Y(s) = -2/(s-3).$

Solving this system of algebraic equations gives

$X(s) = [s - 1 + 7(s-1)/(s-3) - 8/(s-3)]/[(s-1)(s+2) - 4]$

$= [s^2 + 3s - 12]/[(s+3)(s-2)(s-3)]$, and
$Y(s) = -s/[(s+3)(s-2)(s-3)]$.
Using partial fractions gives
$X(s) = A/(s+3) + B/(s-2) + C/(s-3)$, so
$[s^2 + 3s - 12] = A(s-2)(s-3) + B(s+3)(s-3) + C(s+3)(s-2)$.
Setting $s = -3$, 2, and 3 gives (in order) $-12 = A(-5)(-6)$, $-2 = B(-5)$, and $6 = C(6)$, so $A = -2/5$, $B = 2/5$, and $C = 1$.
Thus,
$x(t) = -(2/5)\exp(-3t) + (2/5)\exp 2t + \exp 3t$.
To find $y(t)$, we could either use partial fractions for $Y(s)$ or from the differential equation calculate
$y(t) = [x' + 2x - 7\exp 3t]/4$. Either way gives
$y(t) = (1/10)\exp(-3t) + (2/5)\exp 2t - (1/2)\exp 3t$.

(c) Taking Laplace transforms of both equations and using the result $\mathcal{L}\{y'\} = s\mathcal{L}\{y\} - y(0)$ together with the initial conditions gives
$sX(s) - 1 = 2X(s) + Y(s) + s/(s^2+1)$,
$sY(s) = -3X(s) - 2Y(s)$, or
$(s-2)X(s) - Y(s) = 1 + s/(s^2+1)$,
$3X(s) + (s+2)Y(s) = 0$.
Solving this system of algebraic equations gives
$Y(s) = -3\left(s^2+s+1\right)/[(s-1)(s+1)(s^2+1)]$,
$X(s) = \left(s^2+s+1\right)(s+2)/[(s-1)(s+1)(s^2+1)]$.
Using partial fractions for $Y(s)$ gives
$-3\left(s^2+s+1\right)/[(s-1)(s+1)(s^2+1)] = A/(s-1) + B/(s+1) + (Cs+D)/(s^2+1)$, so
$-3\left(s^2+s+1\right) = A(s+1)(s^2+1) + B(s-1)(s^2+1) + (Cs+D)(s^2-1)$
giving $0 = A+B+C$, $-3 = A-B+D$, $-3 = A+B-C$, and $-3 = A-B-D$.
Thus, $A = -9/4$, $B = 3/4$, $C = 3/2$, and $D = 0$.
$Y(s) = (-9/4)/(s-1) + (3/4)/(s+1) + (3/2)s/(s^2+1)$ and
$y(t) = (-9/4)\exp t + (3/4)\exp(-t) + (3/2)\cos t$.
To evaluate $x(t)$ we could either use partial fractions on $X(s)$ or use $x(t) = [-y' - 2y]/3$. Either method gives
$x(t) = (9/4)\exp t - (1/4)\exp(-t) + (1/2)\sin t - \cos t$.

(e) Taking Laplace transforms of both equations and using the result $\mathcal{L}\{y'\} = s\mathcal{L}\{y\} - y(0)$ together with the initial conditions gives
$sX(s) = -2X(s) + 4Y(s) + \left(e^{-s} - e^{-3s}\right)/s$,
$sY(s) - 1 = X(s) + Y(s)$, or
$(s+2)X(s) - 4Y(s) = \left(e^{-s} - e^{-3s}\right)/s$,
$-X(s) + (s-1)Y(s) = 1$.
Solving this system of algebraic equations gives
$Y(s) = \left[s + 2 + \left(e^{-s} - e^{-3s}\right)/s\right]/[(s+3)(s-2)]$,
$X(s) = 4/[(s+3)(s-2)] + (s-1)\left(e^{-s} - e^{-3s}\right)/[s(s+3)(s-2)]$.
Using partial fractions gives
$(s+2)/[(s+3)(s-2)] = (1/5)/(s+3) + (4/5)/(s-2)$, and
$1/[s(s+3)(s-2)] = (-1/6)/s + (1/15)/(s+3) + (1/10)/(s-2)$.
Thus, $y(t) = (4/5)\exp(2t) + (1/5)\exp(-3t) + f(t-1)\,u(t-1) - f(t-3)\,u(t-3)$, where
$f(t) = -1/6 + (1/15)\exp(-3t) + (1/10)\exp(2t)$.

Similarly,
$4/[(s+3)(s-2)] = (-4/5)/(s+3) + (4/5)/(s-2)$, and
$(s-1)/[s(s+3)(s-2)] = (1/6)/s - (4/15)/(s+3) + (1/10)(s-2)$, so
$x(t) = (4/5)[\exp(2t) - \exp(-3t)] + g(t-1)u(t-1) - g(t-3)u(t-3)$, where
$g(t) = 1/6 - (4/15)\exp(-3t) + (1/10)\exp(2t)$.

7. **(a)** Taking Laplace transforms of both equations and using the result $\mathcal{L}\{y'\} = s\mathcal{L}\{y\} - y(0)$ together with the initial conditions gives
$2sX(s) - 8 - 2X(s) + sY(s) + 3 = 1/s$,
$sX(s) - 4 - 3X(s) + sY(s) + 3 - 3Y(s) = 2/s$, or
$(2s-2)X(s) + sY(s) = 5 + 1/s$,
$(s-3)X(s) + (s-3)Y(s) = 1 + 2/s$.
Solving this system of algebraic equations gives
$X(s) = (4s^2 - 16s - 3)/[s(s-2)(s-3)]$,
$Y(s) = -X(s) + 1/(s-3) + 2/[s(s-3)]$.
Using partial fractions for $X(s)$ gives
$X(s) = (4s^2 - 16s - 3)/[s(s-2)(s-3)] = A/s + B/(s-2) + C/(s-3)$, so
$4s^2 - 16s - 3 = A(s-2)(s-3) + Bs(s-3) + Cs(s-2)$.
Setting $s = 0$, 2, and 3, gives (in order) $-3 = 6A$, $-19 = -2B$, and $-15 = 3C$.
Thus, $A = -1/2$, $B = 19/2$, and $C = -5$, and
$x(t) = -1/2 + (19/2)\exp(2t) - 5\exp(3t)$, and
$y(t) = -x(t) + \exp(3t) - 2/3 + (2/3)\exp(3t)$
$= -1/6 - (19/2)\exp(2t) + (20/3)\exp(3t)$.

(c) Taking Laplace transforms of both equations and using the result $\mathcal{L}\{y'\} = s\mathcal{L}\{y\} - y(0)$ together with the initial conditions gives
$sX(s) + 1 - sY(s) - 10 = -1/(s-1)$,
$2sX(s) + 2 - 2sY(s) - 20 - Y(s) = 8/s$, or
$sX(s) - sY(s) = 9 - 1/(s-1)$,
$2sX(s) - (2s+1)Y(s) = 18 + 8/s$.
Solving this system of algebraic equations gives
$X(s) = 2/s - 3/(s-1)$,
$Y(s) = -8/s - 2/(s-1)$.
Thus,
$x(t) = 2 - 3\exp t$, and
$y(t) = -8 - 2\exp t$.

(e) Taking Laplace transforms of both equations and using the result $\mathcal{L}\{y'\} = s\mathcal{L}\{y\} - y(0)$ together with the initial conditions gives
$s^2X(s) + 4s - 8 + 2X(s) - 4sY(s) + 4 = 0$,
$sX(s) + 4 + s^2Y(s) - s - 2 - 4Y(s) = 0$, or
$(s^2+2)X(s) - 4sY(s) = 4 - 4s$,
$sX(s) + (s^2-4)Y(s) = -2 + s$.
Solving this system of algebraic equations gives
$X(s) = (8-4s)/(s^2+4)$,
$Y(s) = (s+2)/(s^2+4)$.
Thus,
$x(t) = 4\sin 2t - 4\cos 2t$, and
$y(t) = \sin 2t + \cos 2t$.

11.8 When Do Laplace Transforms Exist?

1. Given that $|f(t)| \leq M \exp \alpha t$ for $t > T_1$ and $|g(t)| \leq N \exp \beta t$, for $t > T_2$, then $|f(t)| \leq M \exp \alpha t$ and $|g(t)| \leq N \exp \beta t$, for $t > T$, where $T = \max(T_1, T_2)$.

Now we also have that

$|f(t) + g(t)| \leq |f(t)| + |g(t)| \leq M \exp \alpha t + N \exp \beta t$ for $t > T$.

Let γ be the larger of α and β, then

$|f(t)+g(t)| \leq M \exp \alpha t + N \exp \beta t \leq M \exp \gamma t + N \exp \gamma t = (M+N) \exp \gamma t$, for $t > T$.

This shows that $f(t) + g(t)$ is also of exponential order as t approaches infinity.

3. If $f(t)$ and $g(t)$ are of exponential order as $t \to \infty$ and piecewise continuous, then $\mathcal{L}\{f(t)\}$ and $\mathcal{L}\{g(t)\}$ both exist. From Exercises 1 and 2, $f(t)+g(t)$ is both piecewise continuous and of exponential order as $t \to \infty$. Because of this, and the fact that $\int_0^b [f(x) + g(x)]\, dx = \int_0^b f(x)\, dx + \int_0^b g(x)\, dx$, then $\mathcal{L}\{f(t) + g(t)\} = \mathcal{L}\{f(t)\} + \mathcal{L}\{g(t)\}$.

5. (a) $|\exp 3t \cos 2t| \leq \exp 3t\, |\cos 2t| \leq \exp 3t$ for $t > 0$.
Thus, $\alpha = 3$, $M = 1$, and $T = 0$.

(c) $\exp 2t = 1 + 2t + (2t)^2/2! + \cdots$, so $2t = \exp 2t - 1 - (2t)^2/2! - \cdots \leq \exp 2t$ for $t > 0$.
Thus, $\alpha = 2$, $M = 1$, and $T = 0$.

7. $|f(t)| \leq 3$ so $f(t)$ is of exponential order. $f'(t) = 6t \exp(t^2) \sin(\exp(t^2))$. Because $\lim_{t\to\infty} \exp(t^2)/\exp(\alpha t) = \infty$ for all α, $f'(t)$ is not of exponential order.

9. The function depicted by the graph in Figure 11.28 has corners when $x = 1$ and $x = 3$. At these points the derivative of the function will not exist, therefore this function cannot be a solution of $ay'' + by' + cy = f(t)$.

11. $\mathcal{L}\{f''(t)\} - 2\mathcal{L}\{tf'(t)\} + 2L\{f(t)\} = 0$. Using the properties from Exercise 10 changes this equation to

$s^2 F(s) - sf(0) - f'(0) + 2[sF'(s) + F(s)] + 2F(s) = 0$, or

$2sF'(s) + (s^2 + 4)F(s) = 1$.

Writing this linear differential equation as

$F'(s) + (s/2 + 2/s)F(s) = 1/(2s)$

gives the integrating factor as $s^2 \exp(s^2/4)$.

$d/ds[(s^2 \exp(s^2/4)F(s)] = s/2 \exp(s^2/4)$.

Integration yields

$s^2 \exp(s^2/4)F(s) = \exp(s^2/4) + C$, or

$F(s) = 1/s^2 + C \exp(-s^2/4)/s^2$.

Thus, $f(t) = t + C\mathcal{L}^{-1}\{\exp(-s^2/4)/s^2\}$.

We now notice that $f(t) = t$ satisfies the original differential equation as well as the boundary conditions $f(0) = 0$, $f'(0) = 1$. Theorem 9.1 applies to this situation, the conclusion of which guarantees a unique solution. Because $f(t) = t$ is a solution which satisfies both initial conditions, we must let $C = 0$, and our solution is $f(t) = t$.

12. USING POWER SERIES

12.1 Solutions Using Taylor Series

1. $y'(x) = (x^2 + y^2)^{1/2}$ and $y(0) = 1$ so $y'(0) = 1$.

$y''(x) = (x + yy')(x^2 + y^2)^{-1/2}$, so

$y''(0) = (0 + y(0)y'(0))(0 + y^2(0))^{-1/2} = 1$.

$y'''(x) = -(x + yy')^2 (x^2 + y^2)^{-3/2} + (1 + yy'' + y'y')(x^2 + y^2)^{-1/2}$, so

$y'''(0) = -1 + 3 = 2$.

Thus,

$y(x) = y(0) + y'(0)x + \frac{1}{2!}y''(0)x^2 + \frac{1}{3!}y'''(0)x^3 + \cdots$

$= 1 + x + \frac{1}{2}x^2 + \frac{1}{3}x^3 + \cdots$.

3. $y''(x) = -\sin y$ and $y(0) = 1$, $y'(0) = 0$ so $y''(0) = -\sin 1$.

$y'''(x) = -y' \cos y$, so $y'''(0) = 0$.

$y^{iv}(x) = -y'' \cos y + (y')^2 \sin y$, so $y^{iv}(0) = \sin 1 \cos 1$.

Thus,

$y(x) = y(0) + y'(0)x + \frac{1}{2!}y''(0)x^2 + \frac{1}{3!}y'''(0)x^3 + \frac{1}{4!}y^{iv}(0)x^4 + \cdots$

$= 1 - \left(\frac{1}{2}\sin 1\right)x^2 + \left(\frac{1}{24}\sin 1 \cos 1\right)x^4 + \cdots$.

5. $y'(x) = 2xy$ and $y(0) = 1$ so $y'(0) = 0$.

$y''(x) = 2y + 2xy'$, so $y''(0) = 2$.

$y'''(x) = 4y' + 2xy''$, so $y'''(0) = 0$.

$y^{iv}(x) = 6y'' + 2xy'''$, so $y^{iv}(0) = 6 \cdot 2 = 12 = 3 \cdot 4 = 4!/2!$.

$y^{v}(x) = 8y''' + 2xy^{iv}$, so $y^{v}(0) = 0$.

$y^{vi}(x) = 10y^{iv} + 2xy^{v}$, so $y^{vi}(0) = 10 \cdot 12 = 120 = 4 \cdot 5 \cdot 6 = 6!/3!$.

Thus,

$y(x) = y(0) + y'(0)x + \frac{1}{2!}y''(0)x^2 + \frac{1}{3!}y'''(0)x^3 + \frac{1}{4!}y^{iv}(0)x^4 + \cdots$

$= 1 + x^2 + \frac{1}{2!}x^4 + \frac{1}{3!}x^6 + \cdots$. These are the first 7 terms of the Taylor series for $\exp(x^2)$.

7. $y''(x) = -y$ and $y(0) = y_0$, $y'(0) = y_0^*$, so $y''(0) = -y_0$.

$y'''(x) = -y'$ so $y'''(0) = -y_0^*$.

$y^{iv}(x) = -y''$ so $y^{iv}(0) = y_0$.

In general, $y^{(2n)}(0) = (-1)^n y_0$ and $y^{(2n+1)}(0) = (-1)^n y_0^*$.

Thus,

$y(x) = y(0) + y'(0)x + \frac{1}{2!}y''(0)x^2 + \frac{1}{3!}y'''(0)x^3 + \frac{1}{4!}y^{iv}(0)x^4 + \cdots$

$= y_0\left(1 - \frac{1}{2!}x^2 + \frac{1}{4!}x^4 + \cdots\right) + y_0^*\left(x - \frac{1}{3!}x^3 + \frac{1}{5!}x^5 + \cdots\right)$

$= y_0 \cos x + y_0^* \sin x$.

This is the same answer as the one we get by solving $y'' + y = 0$ — obtaining $y(x) = C_1 \cos x + C_2 \sin x$ — and then imposing the initial conditions $y(0) = y_0$, $y'(0) = y_0^*$ — giving $C_1 = y_0$ and $C_2 = y_0^*$.

12.2 Solutions Using Power Series

1. (a) Set $k = m - 2$, so when $k = 0$, $m = 2$.

   (c) Set $m = k + n$, so when $m = n$, $k = 0$.

3. Both $-x$ and 3 are analytic at $x = 0$, so we may apply the power series method.

   $y(x) = \sum_{n=0}^{\infty} c_n x^n$, $y'(x) = \sum_{n=1}^{\infty} c_n n x^{n-1}$, $y''(x) = \sum_{n=2}^{\infty} c_n n(n-1) x^{n-2}$ gives

   $0 = y'' - xy' + 3y$

   $= \sum_{n=2}^{\infty} c_n n(n-1) x^{n-2} - x \sum_{n=1}^{\infty} c_n n x^{n-1} + 3 \sum_{n=0}^{\infty} c_n x^n$

   $= \sum_{m=0}^{\infty} c_{m+2}(m+2)(m+1) x^m - \sum_{n=1}^{\infty} c_n n x^n + \sum_{n=0}^{\infty} 3c_n x^n$

   $= 2c_2 + 3c_0 + \sum_{m=1}^{\infty} [c_{m+2}(m+2)(m+1) - (m-3)c_m] x^m$.

   Thus, $2c_2 + 3c_0 = 0$ and $c_{m+2}(m+2)(m+1) - (m-3)c_m = 0$ for $m = 1, 2, 3, \cdots$, or

   $c_2 = -\frac{3}{2}c_0$ and $c_{m+2} = \frac{m-3}{(m+2)(m+1)} c_m$ for $m = 1, 2, 3, \cdots$.

   With $m = 1$, $c_3 = \frac{-2}{3 \cdot 2} c_1 = -\frac{1}{3} c_1$.

   With $m = 2$, $c_4 = \frac{-1}{4 \cdot 3} c_2 = -\frac{1}{4 \cdot 3}\left(-\frac{3}{2}c_0\right) = \frac{3}{4!} c_0$.

   With $m = 3$, $c_5 = \frac{0}{5 \cdot 4} c_3 = 0$.

   With $m = 4$, $c_6 = \frac{1}{6 \cdot 5} c_4 = \frac{1}{6 \cdot 5}\left(\frac{3}{4!} c_0\right) = \frac{3}{6!} c_0$.

   With $m = 5$, $c_7 = \frac{2}{7 \cdot 6} c_5 = 0$.

   The series solution is

   $y(x) = c_0 + c_1 x + c_2 x^2 + c_3 x^3 + c_4 x^4 + \cdots$

   $= c_0 + c_1 x - \frac{3}{2} c_0 x^2 - \frac{1}{3} c_1 x^3 + \frac{3}{4!} c_0 x^4 + \frac{3}{6!} c_0 x^6 + \cdots$,

   which can be written

   $y(x) = c_0 \left(1 - \frac{3}{2!} x^2 + \frac{3}{4!} x^4 + \frac{3}{6!} x^6 + \cdots\right) + c_1 \left(x - \frac{1}{3} x^3\right)$.

   To satisfy the initial condition $y(0) = 2$, we find $y(0) = c_0$ so $c_0 = 2$.

   To satisfy the initial condition $y'(0) = 0$, we find

   $y'(x) = c_0 \left(-3x + \frac{1}{2} x^3 + \cdots\right) + c_1 \left(1 - x^2\right)$.

   Thus, $y'(0) = c_1$ so $c_1 = 0$.

   The solution is $y(x) = 2\left(1 - \frac{3}{2!} x^2 + \frac{3}{4!} x^4 + \frac{3}{6!} x^6 + \cdots\right)$.

5. $y = ze^{-x^2} \Longrightarrow y' = (z' - 2xz) e^{-x^2} \Longrightarrow y'' = \left(z'' - 4xz' - 2z + 4x^2 z\right) e^{-x^2}$

   $y'' + 2xy' + 2y =$

   $\left(z'' - 4xz' - 2z + 4x^2 z\right) e^{-x^2} + 2x (z' - 2xz) e^{-x^2} + 2ze^{-x^2} =$

   $(z'' - 2xz') e^{-x^2}$

   Thus, $y'' + 2xy' + 2y = 0$ reduces to

   $z'' - 2xz' = 0$ with solution

   $z' = e^{x^2}$, and so

   $z = \int_0^x e^{t^2} \, dt$.

   Thus,

   $y(x) = e^{-x^2} \int_0^x e^{t^2} \, dt$.

(a) If there was an $x = b > 0$ for which $y(b) = 0$, we would have $e^{-b^2}\int_0^b e^{t^2}\,dt = 0$, or $\int_0^b e^{t^2}\,dt = 0$. However, $e^{t^2} > 0$ so $\int_0^b e^{t^2}\,dt > 0$. Thus, $y(x)$ cannot vanish for $x > 0$.

(b) Extreme values occur at $x = x_m$, where $y'(x_m) = 0$. From $y(x) = e^{-x^2}\int_0^x e^{t^2}\,dt$ we find $y'(x) = 1-2xy(x)$, and so $0 = 1-2x_m y(x_m)$, and $y(x_m) = 1/(2x_m)$. From the differential equation $y''(x) = -2y(x) - 2xy'(x)$, we find $y''(x_m) = -2y(x_m)$. In part (a) we showed that $y(x) > 0$ for $x > 0$, so $y''(x_m) < 0$ if $x_m > 0$. Thus, $x = x_m$ is a maximum. There can be no minimums for $x > 0$, and so there cannot be a second maximum. At the maximum we have $x_m y(x_m) = 1/2$ so $x_m e^{-x_m^2}\int_0^{x_m} e^{t^2}\,dt = 1/2$. By writing this in the form $\int_0^{x_m} x_m e^{t^2-x_m^2}\,dt = 1/2$, or $\int_0^1 x_m^2 e^{(u^2-1)x_m^2}\,du = 1/2$, and experimenting with different choices for x_m we find $x_m \approx 0.924$, in which case $y(x_m) = 1/(2x_m) \approx 0.541$.

(c) From $y'(x) = 1-2xy(x)$ and the differential equation $y''(x) = -2y(x)-2xy'(x)$, we find $y''(x) = -2y(x)-2x\,[1 - 2xy(x)]$, so $y''(x_i) = 0$ when $-2y(x_i)-2x_i\,[1 - 2xy(x_i)] = 0$, or $(2x_i^2 - 1)\,y(x_i) - x_i = 0$, that is, $y(x_i) = x_i/\left(2x_i^2 - 1\right)$. Because we have shown in part (b) that this solution has only one maximum and no minimums it cannot have more than one point of inflection after the maximum. At the point of inflection we have $(2x_i^2 - 1)\,y(x_i)/x_i = 1$ so $(2x_i^2 - 1)\,e^{-x_i^2}/x_i\int_0^{x_i} e^{t^2}\,dt = 1$. By writing this in the form $\int_0^{x_i}(2x_i^2 - 1)\,e^{t^2-x_i^2}/x_i\,dt = 1$, or $\int_0^1 (2x_i^2 - 1)\,e^{(u^2-1)x_i^2}\,du = 1$, and experimenting with different choices for x_i we find $x_i \approx 1.502$, in which case $y(x_i) = x_i/\left(2x_i^2 - 1\right) \approx 0.428$.

(d) $y(x) = e^{-x^2}\int_0^x e^{t^2}\,dt = \int_0^x e^{t^2}\,dt/e^{x^2}$. Both the numerator and denominator go to infinity as $x \to \infty$, so

$$\lim_{x\to\infty} y(x) = \lim_{x\to\infty}\frac{\int_0^x e^{t^2}\,dt}{e^{x^2}} = \lim_{x\to\infty}\frac{\left(\int_0^x e^{t^2}\,dt\right)'}{\left(e^{x^2}\right)'} = \lim_{x\to\infty}\frac{e^{x^2}}{2xe^{x^2}} = 0.$$

(e) $2xy(x) = 2xe^{-x^2}\int_0^x e^{t^2}\,dt = 2x\int_0^x e^{t^2}\,dt/e^{x^2}$. Both the numerator and denominator go to infinity as $x \to \infty$, so

$$\lim_{x\to\infty} 2xy(x) = \lim_{x\to\infty}\frac{2x\int_0^x e^{t^2}\,dt}{e^{x^2}} = \lim_{x\to\infty}\frac{\left(2x\int_0^x e^{t^2}\,dt\right)'}{\left(e^{x^2}\right)'} = \lim_{x\to\infty}\frac{\int_0^x e^{t^2}\,dt + 2xe^{x^2}}{2xe^{x^2}}.$$

But,

$$\lim_{x\to\infty}\frac{\int_0^x e^{t^2}\,dt}{2xe^{x^2}} = \lim_{x\to\infty}\frac{e^{x^2}}{2\,(1+2x)\,e^{x^2}} = 0,$$

so

$$\lim_{x\to\infty} 2xy(x) = 1.$$

(f) The solution is positive for $x > 0$. It starts at the origin, and increases in a concave down manner until it reaches a maximum at $(0.924, 0.541)$. It then decreases but is still concave down until it reaches its point of inflection at $(1.502, 0.428)$. It is then concave up but decreasing, and approaches the horizontal axis asymptotically, like the function $1/(2x)$.

(g) $y(-x) = e^{-(-x)^2}\int_0^{-x} e^{t^2}\,dt = e^{-x^2}\int_0^{-x} e^{t^2}\,dt$. Now let $t = -u$ so that $y(-x) = -e^{-x^2}\int_0^x e^{(-u)^2}\,du = -y(x)$. Thus, the function is odd, and can be sketched for $x < 0$ from $x > 0$.

7. By making the substitution $X = x - 1$, $\frac{d^2y}{dx^2} - 2\,(x-1)\,\frac{dy}{dx} + 2y = 0$ about $x = 1$, becomes $\frac{d^2y}{dX^2} - 2X\frac{dy}{dX} + 2y = 0$ about $X = 0$. Both $-2X$ and 2 are analytic at $X = 0$, so we may apply the power series method.

$y(X)=\sum_{n=0}^{\infty}c_nX^n$, $y'(X)=\sum_{n=1}^{\infty}c_nnX^{n-1}$, $y''(X)=\sum_{n=2}^{\infty}c_nn(n-1)X^{n-2}$ gives

$0=\frac{d^2y}{dX}-2X\frac{dy}{dX}+2y$

$=\sum_{n=2}^{\infty}c_nn(n-1)X^{n-2}-2X\sum_{n=1}^{\infty}c_nnX^{n-1}+2\sum_{n=0}^{\infty}c_nX^n$

$=\sum_{m=0}^{\infty}c_{m+2}(m+2)(m+1)X^m-\sum_{n=1}^{\infty}2c_nnX^n+\sum_{n=0}^{\infty}2c_nX^n$

$=2c_2+2c_0+\sum_{m=1}^{\infty}\left[c_{m+2}(m+2)(m+1)-(2m-2)c_m\right]X^m.$

Thus, $2c_2+2c_0=0$ and $c_{m+2}(m+2)(m+1)-(2m-2)c_m=0$ for $m=1,2,3,\cdots$, or

$c_2=-c_0$ and $c_{m+2}=\frac{2(m-1)}{(m+2)(m+1)}c_m$ for $m=1,2,3,\cdots$.

With $m=1$, $c_3=\frac{0}{3\cdot 2}c_1=0$. From this, $c_5=c_7=c_9=\cdots=0$.

With $m=2$, $c_4=\frac{2}{4\cdot 3}c_2=-\frac{2}{4\cdot 3}c_0$.

With $m=4$, $c_6=\frac{2\cdot 3}{6\cdot 5}c_4=\frac{2\cdot 3}{6\cdot 5}\left(-\frac{2}{4\cdot 3}c_0\right)=-\frac{2^2 3}{6\cdot 5\cdot 4\cdot 3}c_0$.

The series solution is

$y(X)=c_0+c_1X+c_2X^2+c_3X^3+c_4X^4+\cdots$

$=c_0+c_1X-c_0X^2-\frac{2}{4\cdot 3}c_0X^4-\frac{2^2 3}{6\cdot 5\cdot 4\cdot 3}c_0X^6+\cdots$

which can be written

$y(X)=c_0\left(1-X^2-\frac{2}{4\cdot 3}X^4-\frac{2^2 3}{6\cdot 5\cdot 4\cdot 3}X^6+\cdots\right)+c_1X.$

In terms of x this is

$y(x)=c_0\left[1-(x-1)^2-\frac{2}{4\cdot 3}(x-1)^4-\frac{2^2 3}{6\cdot 5\cdot 4\cdot 3}(x-1)^6+\cdots\right]+c_1(x-1).$

9. **(a)** Because x^2, $\sin x$, and 3 are analytic at $x=0$, with radius of convergence ∞, we may apply the power series method and expect the solution to have a radius of convergence ∞.

$y(x)=\sum_{k=0}^{\infty}c_kx^k$, $y'(x)=\sum_{k=1}^{\infty}c_kkx^{k-1}$, $y''(x)=\sum_{k=2}^{\infty}c_kk(k-1)x^{k-2}$ gives

$3=y''+x^2y'+(\sin x)y$

$=\sum_{k=2}^{\infty}c_kk(k-1)x^{k-2}+x^2\sum_{k=1}^{\infty}c_kkx^{k-1}+\sum_{k=0}^{\infty}(-1)^kx^{2k+1}/(2k+1)!\sum_{k=0}^{\infty}c_kx^k$

$=(2c_2+6c_3x+12c_4x^2+\cdots)+x^2(c_1+2c_2x+3c_3x^2\cdots)+(x-x^3/3!+\cdots)(c_0+c_1x+c_2$

$=2c_2+(6c_3+c_0)x+(12c_4+2c_1)x^2+(20c_5+3c_2-c_0/3!)x^3+\cdots.$

Thus, $2c_2=3$, $c_3=-c_0/6$, $c_4=-c_1/6$.

$y(0)=0$ requires that $c_0=0$, and $y'(0)=3$ requires that $c_1=3$. Thus, $c_2=3/2$, $c_3=0$, and $c_4=-1/2$, so that

$y(x)=3x+\frac{3}{2}x^2-\frac{1}{2}x^4+\cdots.$ $R=\infty$.

(c) Both $-2x$ and e^x are analytic at $x=0$, with radius of convergence ∞, so we may apply the power series method and expect the solution to have a radius of convergence ∞.

$y(x)=\sum_{k=0}^{\infty}c_kx^k$, $y'(x)=\sum_{k=1}^{\infty}c_kkx^{k-1}$, $y''(x)=\sum_{k=2}^{\infty}c_kk(k-1)x^{k-2}$ gives

$e^x=y''-2xy'+e^xy$

$=\sum_{k=2}^{\infty}c_kk(k-1)x^{k-2}-2x\sum_{k=1}^{\infty}c_kkx^{k-1}+\sum_{k=0}^{\infty}(-1)^kx^k/k!\sum_{k=0}^{\infty}c_kx^k$

$=(2c_2+6c_3x+12c_4x^2+\cdots)-2x(c_1+2c_2x+\cdots)+(1+x+x^2/2!+\cdots)(c_0+c_1x+c_2x$

$=(2c_2+c_0)+(6c_3-c_1+c_0)x+(12c_4-3c_2+c_1)x^2+\cdots.$

Because $e^x=1+x+x^2/2!+\cdots$ we find

$2c_2+c_0=1$, $6c_3-c_1+c_0=1$, $12c_4-3c_2+c_1=1/2$, so that

$c_2=(1-c_0)/2$, $c_3=(1+c_1-c_0)/6$, $c_4=(1/2+3c_2-c_1)/12$.

$y(0)=0$ requires that $c_0=0$, and $y'(0)=-2$ requires that $c_1=-2$. Thus, $c_2=1/2$, $c_3=-1/6$, and $c_4=1/3$, so that

$y(x)=-2x+\frac{1}{2}x^2-\frac{1}{6}x^3+\frac{1}{3}x^4+\cdots.$ $R=\infty$.

(e) Because the initial conditions are given at $x = 1$, then $x_0 = 1$.

By making the substitution $X = x - 1$, $\frac{d^2y}{dx^2} - x\frac{dy}{dx} + (\ln x)\, y = 0$ about $x = 1$, becomes $\frac{d^2y}{dX} - (X+1)\frac{dy}{dX} + \ln(1+X)\, y = 0$ about $X = 0$. Both $-(X+1)$ and $\ln(1+X)$ are analytic at $X = 0$, with radius of convergence ∞ and 1, respectively, so we may apply the power series method and expect the solution to have at least a radius of convergence 1.

$y(X) = \sum_{n=0}^{\infty} c_n X^n$, $y'(X) = \sum_{n=1}^{\infty} c_n n X^{n-1}$, $y''(X) = \sum_{n=2}^{\infty} c_n n(n-1) X^{n-2}$ gives

$0 = \frac{d^2y}{dX} - (X+1)\frac{dy}{dX} + \ln(1+X)\, y$
$= (2c_2 + 6c_3X + 12c_4X^2 + \cdots) +$
$\quad -(X+1)(c_1 + 2c_2X + 3c_3X^2 + \cdots) +$
$\quad +(X + X^2/2 + \cdots)(c_0 + c_1X + c_2X^2 + \cdots)$
$= (2c_2 - c_1) + (6c_3 - c_1 - 2c_2 + c_0)X + (12c_4 - 2c_2 - 3c_3 + c_1 + c_0/2)X^2 + \cdots$.

This gives $c_2 = c_1/2$, $c_3 = (c_1 + 2c_2 - c_0)/6$, $c_4 = (2c_2 + 3c_3 - c_1 - c_0/2)/12$.

The initial conditions, in terms of X are $y(0) = 0$ and $y'(0) = 2$ so that $c_0 = 0$, and $c_1 = 2$.

Thus, $c_2 = 1$, $c_3 = 2/3$, $c_4 = 1/6$, giving
$y(X) = 2X + X^2 + \frac{2}{3}X^3 + \frac{1}{6}X^4 + \cdots$, or
$y(x) = 2(x-1) + (x-1)^2 + \frac{2}{3}(x-1)^3 + \frac{1}{6}(x-1)^4 + \cdots$. $R = 1$.

11. (a) Both 0 and x^2 are analytic at $x = 0$, with radius of convergence ∞, so we may apply the power series method and expect the solution to have a radius of convergence ∞.

$y(x) = \sum_{k=0}^{\infty} c_k x^k$, $y'(x) = \sum_{k=1}^{\infty} c_k k x^{k-1}$, $y''(x) = \sum_{k=2}^{\infty} c_k k(k-1) x^{k-2}$ gives

$0 = y'' + x^2 y$
$= \sum_{k=2}^{\infty} c_k k(k-1) x^{k-2} + x^2 \sum_{k=0}^{\infty} c_k x^k$
$= \sum_{k=-2}^{\infty} c_{k+4}(k+4)(k+3) x^{k+2} + \sum_{k=0}^{\infty} c_k x^{k+2}$
$= 2 \cdot 1c_2 + 3 \cdot 2c_3 x + \sum_{k=0}^{\infty} [c_{k+4}(k+4)(k+3) + c_k] x^{k+2}$.

Thus, $c_2 = 0$, $c_3 = 0$, and $c_{k+4}(k+4)(k+3) + c_k = 0$ for $k = 0, 1, 2, \cdots$.

This gives $c_4 = -c_0/(4 \cdot 3)$, $c_5 = -c_1/(5 \cdot 4)$, $c_6 = -c_2/(6 \cdot 5) = 0$, $c_7 = -c_3/(7 \cdot 6) = 0$, $c_8 = -c_4/(8 \cdot 7) = (-1)^2 c_0/(8 \cdot 7 \cdot 4 \cdot 3)$, $c_9 = -c_5/(9 \cdot 8) = (-1)^2 c_1/(9 \cdot 8 \cdot 5 \cdot 4)$, and so on.

Thus, $y(x) = c_0\left[1 - \frac{1}{4\cdot 3}x^4 + (-1)^2 \frac{1}{8\cdot 7\cdot 4\cdot 3}x^8 + \cdots\right] + c_1\left[x - \frac{1}{5\cdot 4}x^5 + (-1)^2 \frac{1}{9\cdot 8\cdot 5\cdot 4}x^9 + \cdots\right]$.
$R = \infty$.

(c) Both $2x/(1-x^2)$ and $5/(1-x^2)$ are analytic at $x = 0$, with radius of convergence 1, so we may apply the power series method and expect the solution to have at least a radius of convergence 1.

$y(x) = \sum_{k=0}^{\infty} c_k x^k$, $y'(x) = \sum_{k=1}^{\infty} c_k k x^{k-1}$, $y''(x) = \sum_{k=2}^{\infty} c_k k(k-1) x^{k-2}$ gives

$0 = (1-x^2) y'' + 2xy' + 5y$
$= (1-x^2)\sum_{k=2}^{\infty} c_k k(k-1) x^{k-2} + 2x\sum_{k=1}^{\infty} c_k k x^{k-1} + 5\sum_{k=0}^{\infty} c_k x^k$
$= \sum_{k=2}^{\infty} c_k k(k-1) x^{k-2} - \sum_{k=2}^{\infty} c_k k(k-1) x^k + \sum_{k=1}^{\infty} 2c_k k x^k + \sum_{k=0}^{\infty} 5c_k x^k$
$= \sum_{k=0}^{\infty} c_{k+2}(k+2)(k+1) x^k - \sum_{k=2}^{\infty} c_k k(k-1) x^k + \sum_{k=1}^{\infty} 2c_k k x^k + \sum_{k=0}^{\infty} 5c_k x^k$
$= 2 \cdot 1c_2 + 3 \cdot 2c_3 x + \sum_{k=2}^{\infty} c_{k+2}(k+2)(k+1) x^k - \sum_{k=2}^{\infty} c_k k(k-1) x^k +$
$\quad + 2c_1 x + \sum_{k=2}^{\infty} 2c_k k x^k + 5c_0 + 5c_1 x + \sum_{k=2}^{\infty} 5c_k x^k$
$= (2 \cdot 1c_2 + 5c_0) + (3 \cdot 2c_3 + 7c_1)x + \sum_{k=2}^{\infty} [c_{k+2}(k+2)(k+1) - c_k k(k-1) + 2c_k k + 5c_k] x^k$
$= (2 \cdot 1c_2 + 5c_0) + (3 \cdot 2c_3 + 7c_1)x + \sum_{k=2}^{\infty} [c_{k+2}(k+2)(k+1) - c_k(k^2 - 3k - 5)] x^k$.

Thus, $2 \cdot 1c_2 + 5c_0 = 0$, $3 \cdot 2c_3 + 7c_1 = 0$, and $c_{k+2}(k+2)(k+1) - c_k(k^2 - 3k - 5) = 0$ for $k = 2, 3, 4, \cdots$.

This gives $c_2 = -(5/2)\,c_0$, $c_3 = -(7/6)\,c_1$, and $c_{k+2} = c_k\left(k^2 - 3k - 5\right)/(k+2)(k+$ for $k = 2, 3, 4, \cdots$, so

$c_4 = (-7/12)\,c_2 = (35/24)\,c_0$, $c_5 = (-5/20)\,c_3 = -c_3/4 = (7/24)\,c_1$.

Thus, $y(x) = c_0 + c_1 x - \frac{5}{2}c_0 x^2 - \frac{7}{6}c_1 x^3 + \frac{35}{24}c_0 x^4 + \frac{7}{24}c_1 x^5 + \cdots$. $R = 1$.

13. Both $-2x/\left(1-x^2\right)$ and $\lambda/\left(1-x^2\right)$ are analytic at $x = 0$, with radius of convergence 1, so we may apply the power series method and expect the solution to have at least a radius of convergence 1.

$y(x) = \sum_{k=0}^{\infty} c_k x^k$, $y'(x) = \sum_{k=1}^{\infty} c_k k x^{k-1}$, $y''(x) = \sum_{k=2}^{\infty} c_k k\,(k-1)\,x^{k-2}$ gives

$0 = \left(1-x^2\right)y'' - 2xy' + \lambda y$

$= \left(1-x^2\right)\sum_{k=2}^{\infty} c_k k\,(k-1)\,x^{k-2} - 2x\sum_{k=1}^{\infty} c_k k x^{k-1} + \lambda\sum_{k=0}^{\infty} c_k x^k$

$= \sum_{k=2}^{\infty} c_k k\,(k-1)\,x^{k-2} - \sum_{k=2}^{\infty} c_k k\,(k-1)\,x^k - \sum_{k=1}^{\infty} c_k 2k x^k + \sum_{k=0}^{\infty} c_k \lambda x^k$

$= \sum_{m=0}^{\infty} c_{m+2}\,(m+2)\,(m+1)\,x^m - \sum_{k=2}^{\infty} c_k k\,(k-1)\,x^k - \sum_{k=1}^{\infty} c_k 2k x^k + \sum_{k=0}^{\infty} c_k \lambda x^k$

$= 2c_2 + \lambda c_0 + [3\cdot 2c_3 - (2-\lambda)\,c_1]\,x + \sum_{m=2}^{\infty}\{c_{m+2}\,(m+2)\,(m+1) - c_m\,[m\,(m-1) + 2m -$

$= 2c_2 + \lambda c_0 + [3\cdot 2c_3 - (2-\lambda)\,c_1]\,x + \sum_{m=2}^{\infty}\{c_{m+2}\,(m+2)\,(m+1) - c_m\,[m\,(m+1) - \lambda]\}\,x^{*}$

Thus, $2c_2 + \lambda c_0 = 0$, $3\cdot 2c_3 - (2-\lambda)\,c_1 = 0$ and $c_{m+2}\,(m+2)\,(m+1) - c_m\,[m\,(m+1) - \lambda] = 0$ for $m = 2, 3, 4, \cdots$.

(a) If $\lambda = n\,(n+1)$ then $2c_2 + \lambda c_0 = 0$, $3\cdot 2c_3 - (1-n)\,(2+n)\,c_1 = 0$ and $c_{m+2}\,(m+2)\,(m+1) - c_m\,(m-n)\,(m+n+1) = 0$ for $m = 2, 3, 4, \cdots$, or $c_2 = -\frac{1}{2}n\,(n+1)\,c_0$, $c_3 = \frac{1}{3\cdot 2}(1-n)\,(2+n)\,c_1 = 0$ and $c_{m+2} = \frac{(m-n)(m+n+1)}{(m+2)(m+1)}c_m$ for $m = 2, 3, 4, \cdots$.

If n is an even integer then from the recurrence relation with $m = n$ we have $c_{n+2} = 0$. By reusing the recurrence relation. with $m = n+2$ and so on, we see that all the c_m, where m is even, will vanish after c_n. For example, if $n = 2$ then $c_4 = c_6 = \cdots = 0$. This means that the coefficient of c_0 will be a polynomial of degree n. Similar arguments apply if n is odd.

(c) $w(x) = \left(x^2-1\right)^m$, so $w' = 2mx\left(x^2-1\right)^{m-1}$, and $\left(x^2-1\right)w' = 2mx\left(x^2-1\right)^m = 2mxw$, or $\left(1-x^2\right)w' + 2mxw = 0$.

Using Liebnitz rule

$$\frac{d^n}{dx^n}(fg) = \sum_{k=0}^{n}\frac{n!}{(n-k)!k!}\frac{d^{n-k}}{dx^{n-k}}(f)\,\frac{d^k}{dx^k}(g)$$

with $n = m+1$, $f(x) = 1-x^2$, and $g(x) = w'$, noting that $\frac{d^{n-k}}{dx^{n-k}}(f)$ is zero unless $k = 0$, 1, or 2, gives

$$\frac{d^{m+1}}{dx^{m+1}}\left[\left(1-x^2\right)w'\right] = \left(1-x^2\right)\frac{d^{m+2}w}{dx^{m+2}} - 2\,(m+1)\,x\frac{d^{m+1}w}{dx^{m+1}} - m\,(m+1)\,\frac{d^m w}{dx^m}.$$

Using Liebnitz rule with $n = m+1$, $f(x) = x$, and $g(x) = w$, noting that $\frac{d^{n-k}}{dx^{n-k}}(f)$ is zero unless $k = 0$ or 1, gives

$$\frac{d^{m+1}}{dx^{m+1}}(xw) = x\frac{d^{m+1}w}{dx^{m+1}} + 2m\,(m+1)\,\frac{d^m w}{dx^m}.$$

Thus,

$$0 = \frac{d^{m+1}}{dx^{m+1}}\left[\left(1-x^2\right)w' + 2mxw\right] = \left(1-x^2\right)\frac{d^{m+2}w}{dx^{m+2}} - 2x\frac{d^{m+1}w}{dx^{m+1}} + m\,(m+1)\,\frac{d^m w}{dx^m}.$$

This can be written

$$\left(1-x^2\right)\frac{d^2}{dx^2}\left(\frac{d^m w}{dx^m}\right)-2x\frac{d}{dx}\left(\frac{d^m w}{dx^m}\right)+m\left(m+1\right)\left(\frac{d^m w}{dx^m}\right)=0$$

so, $\frac{d^m w}{dx^m}$ is a solution of Legendre's equation, as is $c\frac{d^m w}{dx^m}$, where c is any constant .

15. **(a)** The singular points will occur where $1-x=0$, $x+2=0$, or $x+3=0$, that is, 1, -2, -3.

(c) The singular points will occur where $x+6=0$ or $x-1=0$, that is, -6, 1.

(e) The singular points will occur where $x^2+9=0$ or $x+7=0$, that is, $3i$, $-3i$, -7.

(g) The singular points will occur where $\sin x=0$, $x-1=0$, or $x+2=0$, that is, 1, -2, $\pm n\pi$, $n=0,1,2,\cdots$.

12.3 What To Do When Power Series Fail

1. **(a)** Regular 0, -1, 4, -4. Irregular 1.

(c) Regular -1, $\pm n\pi$, $n=1,2,3,\cdots$. Irregular 0.

(e) Regular -2, 4, -4. Irregular 0.

(g) Regular 1. Irregular 0, 4.

3. **(a)** $p(x)=2(1+x)/(4x)$, $q(x)=1/(4x)$ so
$xp(x)=(1+x)/2$,
$x^2q(x)=4x$.
Thus, $x=0$ is a regular singular point and by the Regular Singular Point Theorem the accompanying series converges for all x.
Substituting $y=x^s$ into the left-hand side of the differential equation gives
$(4x)s(s-1)x^{s-2}+2(1+x)sx^{s-1}+x^s$ or
$4s(s-1)x^{s-1}+2sx^{s-1}+2sx^s+x^s$.
Thus, the indicial equation is $4s^2-2s=0$.

(c) $p(x)=2/(4x)$, $q(x)=1/(4x)$ so
$xp(x)=1/2$,
$x^2q(x)=4x$.
Thus, $x=0$ is a regular singular point and by Regular Singular Point Theorem the accompanying series converges for all x.
Substituting $y=x^s$ into the left-hand side of the differential equation gives
$(4x)s(s-1)x^{s-2}+(2)sx^{s-1}+x^s$ or
$4s(s-1)x^{s-1}+2sx^{s-1}+x^s$.
Thus, the indicial equation is $4s^2-2s=0$.

(e) $p(x)=-1/(2x)$, $q(x)=(x-5)/(2x^2)$ so
$xp(x)=-1/2$,
$x^2q(x)=(x-5)/2$.
Thus, $x=0$ is a regular singular point and by Regular Singular Point Theorem the accompanying series converges for all x.
Substituting $y=x^s$ into the left-hand side of the differential equation gives
$(2x^2)s(s-1)x^{s-2}-(x)sx^{s-1}+(x-5)x^s$ or
$2s(s-1)x^s-sx^s-5x^{s+1}-5x^s$.
Thus, the indicial equation is $2s^2-3s-5=0$.

(g) $p(x) = -3(x+2)/[2(x+1)]$, $q(x) = (3x+5)/[2(x+1)^2]$ so
$(x+1)p(x) = 3(x+2)/2,$
$(x+1)^2q(x) = (3x+5)/2.$
Thus, $x = -1$ is a regular singular point and by Regular Singular Point Theorem the accompanying series converges for all x.
Substituting $y = (x+1)^s$ into the left-hand side of the differential equation gives
$2(x+1)^2s(s-1)(x+1)^{s-2} - 3(x+2)(x+1)s(x+1)^{s-1} + (3x+5)(x+1)^s$ or
$2s(s-1)(x+1)^s - 3s[(x+1)+1](x+1)^s + (3x+3+2)(x+1)^s.$
Thus, the indicial equation is $2s(s-1) - 3s + 2 = 0.$

(i) $p(x) = 1/[3(x-1)]$, $q(x) = -1/[3(x-1)]$ so
$(x-1)p(x) = 1/3,$
$(x-1)^2q(x) = -(x-1)/3.$
Thus, $x = 1$ is a regular singular point and by Regular Singular Point Theorem the accompanying series converges for all x.
Substituting $y = (x-1)^s$ into the left hand side of the differential equation gives
$3(x-1)s(s-1)(x-1)^{s-2} + s(x-1)^{s-1} - (x-1)^s$ or
$3s(s-1)(x-1)^{s-1} + s(x-1)^{s-1} - (x-1)^s.$
Thus, the indicial equation is $3s(s-1) + s = 0.$

(k) $p(x) = (3x^2-4x+1)/(x-1)^2 = (3x-1)(x-1)/(x-1)^2 = (3x-1)/(x-1)$,
$q(x) = -2/(x-1)^2$ so
$(x-1)p(x) = 3x-1,$
$(x-1)^2q(x) = -2.$
Thus, $x = 1$ is a regular singular point and by Regular Singular Point Theorem the accompanying series converges for all x.
Substituting $y = (x-1)^s$ into the left hand side of the differential equation gives
$(x-1)^2s(s-1)(x-1)^{s-2} + (3x-1)(x-1)s(x-1)^{s-1} - 2(x-1)^s$ or
$s(s-1)(x-1)^s + [3(x-1)+2](x-1)s(x-1)^{s-1} - 2(x-1)^s$, or
$[s(s-1)+2s-2](x-1)^s + 3s(x-1)^{s+1}.$
Thus, the indicial equation is $s(s-1) + 2s - 2 = 0.$

12.4 Solutions Using The Method of Frobenius

1. (a) If we substitute $y = x^s$ into the left-hand side of the differential equation we obtain
$s(s-1)x^{s-1} - (3+x)sx^{s-1} + 2x^s$
$= [s(s-1) - 3s]x^{s-1} + (-s+2)x^s$
$= (s^2-4s)x^{s-1} + (-s+2)x^s,$
so the indicial equation is $s^2 - 4s = 0$, with roots $s = 0$ and $s = 4$.
To find the recurrence relation we use $y(x) = \sum_{n=0}^{\infty} c_n x^{n+s}$, $y'(x) = \sum_{n=0}^{\infty}(n+s)c_n x^{n+s-1}$ $y''(x) = \sum_{n=0}^{\infty}(n+s)(n+s-1)c_n x^{n+s-2}$. Substituting these into the differential equation yields

$$x\sum_{n=0}^{\infty}(n+s)(n+s-1)c_n x^{n+s-2} + (-3-x)\sum_{n=0}^{\infty}(n+s)c_n x^{n+s-1} + 2\sum_{n=0}^{\infty}c_n x^{n+s} = 0$$

$$\sum_{n=0}^{\infty}(n+s)(n+s-1)c_n x^{n+s-1}+\sum_{n=0}^{\infty}\left[(-3)(n+s)c_n x^{n+s-1}-(n+s)c_n x^{n+s}\right]+$$

$$+\sum_{n=0}^{\infty}2c_n x^{n+s}=0$$

$$\sum_{n=0}^{\infty}\left[(n+s)(n+s-1)-3(n+s)\right]c_n x^{n+s-1}+\sum_{n=0}^{\infty}\left[-(n+s)+2\right]c_n x^{n+s}=0$$

$$[s(s-1)-3s]\,c_0 x^{s-1}+\sum_{n=1}^{\infty}\left[(n+s)(n+s-1)-3(n+s)\right]c_n x^{n+s-1}+$$

$$+\sum_{n=1}^{\infty}\left[-(n-1+s)+2\right]c_{n-1}x^{n+s-1}=0$$

$$[s(s-1)-3s]\,c_0 x^{s-1}+$$

$$+\sum_{n=1}^{\infty}\left\{\left[(n+s)(n+s-1)-3(n+s)\right]c_n+\left[-(n-1+s)+2\right]c_{n-1}\right\}x^{n+s-1}=0.$$

We want each of the coefficients to be zero when s takes on the values of 0 or 4.

$$\left[(n+s)(n+s-1)-3(n+s)\right]c_n+\left[-(n-1+s)+2\right]c_{n-1}=0, \qquad n=1,2,3,\cdots.$$

If we let $s=0$ the recurrence relation becomes

$$[n(n-1)-3n]\,c_n+[-(n-1)+2]\,c_{n-1}=0, \qquad n=1,2,3,\cdots,$$

or

$$n(n-4)\,c_n=(n-3)\,c_{n-1}, \qquad n=1,2,3,\cdots.$$

Thus, c_0 is arbitrary,

$1(-3)c_1=(-2)c_0$, or $c_1=(2/3)c_0$,

$2(-2)c_2=(-1)c_1$, or $c_2=(1/4)c_1=[2/(4\cdot 3)]c_0$,

$3(-1)c_3=(0)c_2$, or $c_3=0$,

$(0)c_4=(1)c_3=0$, so c_4 is arbitrary.

$c_n=\{(n-3)/[n(n-4)]\}\,c_{n-1}$, $n=5,6,7,\cdots$.

$c_5=[2/(5\cdot 1)]\,c_4$.

$c_6=[3/(6\cdot 2)]\,c_5=[(3\cdot 2)/(6\cdot 5\cdot 2\cdot 1)]\,c_4=[(3!4!)/(6!2!)]\,c_4$.

$c_7=[4/(7\cdot 3)]\,c_6=[(4\cdot 3\cdot 2)/(7\cdot 6\cdot 5\cdot 3\cdot 2\cdot 1)]\,c_4=[(4!4!)/(7!3!)]\,c_4$.

$c_n=[(n-3)!4!/(n!(n-4)!)]\,c_4=[(n-3)\,4!/(n!)]\,c_4$ and the general solution is

$$y(x)=c_0\left(1+\frac{2}{3}x+\frac{1}{6}x^2\right)+c_4\sum_{n=4}^{\infty}\frac{(n-3)4!}{n!}x^n.$$

1. (c) If we substitute $y=x^s$ into the left-hand side of the differential equation and simplify we obtain $\left(s^2-2s\right)x^{s-1}+(s+1)\,x^{s+2}$, so the indicial equation is $s^2-2s=0$, with roots $s=0$ and $s=2$.

To find the recurrence relation we use $y(x)=\sum_{n=0}^{\infty}c_n x^{n+s}$, $y'(x)=\sum_{n=0}^{\infty}(n+s)\,c_n x^{n+s-1}$, $y''(x)=\sum_{n=0}^{\infty}(n+s)(n+s-1)\,c_n x^{n+s-2}$. Substituting these into the differential equation yields

$$\sum_{n=0}^{\infty}(n+s)(n+s-1)\,c_n x^{n+s-1}+\sum_{n=0}^{\infty}(n+s)\,c_n x^{n+s+2}+$$

$$-\sum_{n=0}^{\infty}(n+s)\,c_n x^{n+s-1}+\sum_{n=0}^{\infty}c_n x^{n+s+2}=0$$

$$\sum_{n=0}^{\infty}[(n+s)\,(n+s-1)-(n+s)]\,c_n x^{n+s-1}+\sum_{n=0}^{\infty}[(n+s)+1]\,c_n x^{n+s+2}=0$$

$$\sum_{n=0}^{\infty}(n+s)\,(n+s-2)\,c_n x^{n+s-1}+\sum_{n=3}^{\infty}[(n-3+s)+1]\,c_{n-3}x^{n+s-1}=0$$

$$s(s-2)c_0x^{s-1}+(1+s)\,(s-1)\,c_1x^s+(2+s)\,sc_2x^{s+1}+$$

$$+\sum_{n=3}^{\infty}[(n+s)\,(n+s-2)\,c_n+(n+s-2)\,c_{n-3}]\,x^{n+s-1}=0.$$

We want each of the coefficients to be zero when s takes on the values of 0 or 2.

$$(1+s)\,(s-1)\,c_1=0$$

$$(2+s)\,sc_2=0$$

$$(n+s-2)\,[(n+s)\,c_n+c_{n-3}]=0, \qquad n=3,4,5,\cdots.$$

If $s=0$ then $c_1=0$, c_2 is arbitrary, and

$$nc_n=-c_{n-3}\Longrightarrow c_n=\frac{-1}{n}c_{n-3}, \qquad n=3,4,5,\cdots.$$

$c_4=c_7=c_{10}=c_{13}=\cdots=0.$
$c_3=-c_0/3,$
$c_5=-c_2/5,$
$c_6=-c_3/6=c_0/(6\cdot 3)=c_0/[3^2(2\cdot 1)],$
$c_8=-c_5/8=c_2/(8\cdot 5),$
$c_9=-c_6/9=-c_0/(9\cdot 6\cdot 3)=-c_0[3^3(3\cdot 2\cdot 1)],$
$c_{11}=-c_8/11=-c_2/(11\cdot 8\cdot 5)$. In general

$$c_{3n}=\frac{(-1)^n}{3^n n!}c_0, \qquad n=1,2,3,\cdots$$

$$c_{3n-1}=\frac{(-1)^{n+1}}{(3n-1)(3n-4)\cdots(8)(5)}c_2, \qquad n=2,3,4,\cdots$$

so, because $c_1=c_4=c_7=c_{10}=c_{13}=\cdots=0$, the general solution is

$$y(x)=\sum_{n=0}^{\infty}c_n x^{n+0}=\sum_{n=0}^{\infty}c_{3n}x^{3n}+\sum_{n=1}^{\infty}c_{3n-1}x^{3n-1}$$

or

$$y(x)=c_0\sum_{n=0}^{\infty}\frac{(-1)^n}{3^n n!}x^{3n}+c_2\left[x^2+\sum_{n=2}^{\infty}\frac{(-1)^{n+1}}{(3n-1)(3n-4)\cdots(8)(5)}x^{3n-1}\right].$$

Note:

$$\sum_{n=0}^{\infty}\frac{(-1)^n}{3^n n!}x^{3n}=e^{-\frac{1}{3}x^3}.$$

1. (e) If we substitute $y = x^s$ into the left-hand side of the differential equation $2xy'' + (2x+1)y' + 2y = 0$ and simplify we obtain $[2s(s-1)+s]x^{s-1} + (2s+2)x^s$, so the indicial equation is $2s(s-1)+s = 0$, with roots $s = 0$ and $s = 1/2$.

To find the recurrence relation we use $y(x) = \sum_{n=0}^{\infty} c_n x^{n+s}$, $y'(x) = \sum_{n=0}^{\infty}(n+s)c_n x^{n+s-1}$, $y''(x) = \sum_{n=0}^{\infty}(n+s)(n+s-1)c_n x^{n+s-2}$. Substituting these into the differential equation yields

$$\sum_{n=0}^{\infty} 2(n+s)(n+s-1)c_n x^{n+s-1} + \sum_{n=0}^{\infty} 2(n+s)c_n x^{n+s} +$$

$$+\sum_{n=0}^{\infty}(n+s)c_n x^{n+s-1} + \sum_{n=0}^{\infty} 2c_n x^{n+s} = 0$$

$$[2s(s-1)+s]c_0 x^{s-1} + \sum_{n=0}^{\infty}[(n+1+s)(2n+2s+1)c_{n+1} + 2(n+s+1)c_n]x^{n+s} = 0.$$

We want each of the coefficients to be zero when s takes on the values of 0 or 1/2.

$$(n+1+s)[(2n+2s+1)c_{n+1} + 2c_n] = 0, \qquad n = 0, 1, 2, \cdots.$$

If $s = 0$ then $(2n+1)c_{n+1} + 2c_n = 0$, and

$$c_n = \frac{(-1)^n 2^n}{(2n-1)(2n-3)\cdots(3)(1)}c_0, \qquad n = 1, 2, 3, \cdots.$$

If $s = 1/2$ then $(2n+2)c_{n+1} + 2c_n = 0$, and

$$c_n = \frac{(-1)^n}{n!}c_0, \qquad n = 1, 2, 3, \cdots.$$

Thus,

$$y(x) = c_0\left[1 + \sum_{n=1}^{\infty}\frac{(-1)^n(2x)^n}{(2n-1)(2n-3)\cdots(3)(1)}\right] + c_0^* x^{1/2}\sum_{n=0}^{\infty}\frac{(-1)^n x^n}{n!}.$$

1. (g) If we substitute $y = x^s$ into the left-hand side of the differential equation $2x^2y'' - xy' + (x-5)y = 0$ and simplify we obtain $[2s(s-1) - s - 5] + x^{s+1}$, so the indicial equation is $2s(s-1) - s - 5 = 0$, with roots $s = -1$ and $s = 5/2$.

To find the recurrence relation we use $y(x) = \sum_{n=0}^{\infty} c_n x^{n+s}$, $y'(x) = \sum_{n=0}^{\infty}(n+s)c_n x^{n+s-1}$, $y''(x) = \sum_{n=0}^{\infty}(n+s)(n+s-1)c_n x^{n+s-2}$. Substituting these into the differential equation yields

$$\sum_{n=0}^{\infty} 2(n+s)(n+s-1)c_n x^{n+s} - \sum_{n=0}^{\infty}(n+s)c_n x^{n+s} +$$

$$+\sum_{n=0}^{\infty} c_n x^{n+s+1} - \sum_{n=0}^{\infty} 5c_n x^{n+s} = 0.$$

$$[2s(s-1) - s - 5]c_0 x^s + \sum_{n=1}^{\infty}[(2n+2s-5)(n+s+1)c_n + c_{n-1}]x^{n+s} = 0.$$

We want each of the coefficients to be zero when s takes on the values of -1 or $5/2$.

$$(2n+2s-5)(n+s+1)c_n + c_{n-1} = 0, \qquad n = 1, 2, 3, \cdots.$$

If $s=-1$ then $(2n-7)\,nc_n + c_{n-1} = 0$, and

$$c_n = \frac{(-1)^n}{n!\,(2n-7)\,(2n-9)\cdots(-3)\,(-5)}c_0, \qquad n = 1,2,3,\cdots.$$

If $s=5/2$ then $n\,(2n+7)\,c_n + c_{n-1} = 0$, and

$$c_n = \frac{(-1)^n}{n!\,(2n+7)\,(2n+5)\cdots(11)\,(9)}c_0, \qquad n = 1,2,3,\cdots.$$

Thus,

$$y(x) = c_0 x^{-1}\left[1 + \sum_{n=1}^{\infty}\frac{(-1)^n\,x^n}{n!\,(2n-7)\,(2n-9)\cdots(-3)\,(-5)}\right] +$$

$$+c_0^* x^{5/2}\left[1 + \sum_{n=1}^{\infty}\frac{(-1)^n\,x^n}{n!\,(2n+7)\,(2n+5)\cdots(11)\,(9)}\right].$$

3. **(a)** $y(x) = \sum_{n=0}^{\infty} c_n\,(x-1)^{n+s}$,
$y'(x) = \sum_{n=0}^{\infty}(n+s)\,c_n\,(x-1)^{n+s-1}$,
$y''(x) = \sum_{n=0}^{\infty}(n+s)\,(n+s-1)\,c_n\,(x-1)^{n+s-2}$.
Substituting these into the differential equation yields

$$2\,(x-1)^2\sum_{n=0}^{\infty}(n+s)\,(n+s-1)\,c_n\,(x-1)^{n+s-2} +$$

$$+5\,(x-1)\sum_{n=0}^{\infty}(n+s)\,c_n\,(x-1)^{n+s-1} + [(x-1)+1]\sum_{n=0}^{\infty}c_n\,(x-1)^{n+s} = 0$$

$$2\sum_{n=0}^{\infty}(n+s)\,(n+s-1)\,c_n\,(x-1)^{n+s} + 5\sum_{n=0}^{\infty}(n+s)\,c_n\,(x-1)^{n+s} +$$

$$+\sum_{n=0}^{\infty}c_n\,(x-1)^{n+s+1} + \sum_{n=0}^{\infty}c_n\,(x-1)^{n+s} = 0$$

$$\sum_{n=0}^{\infty}[2\,(n+s)\,(n+s-1) + 5\,(n+s) + 1]\,c_n\,(x-1)^{n+s} + \sum_{n=0}^{\infty}c_n\,(x-1)^{n+s+1} = 0$$

$$\sum_{n=0}^{\infty}[2\,(n+s)\,(n+s) + 3\,(n+s) + 1]\,c_n\,(x-1)^{n+s} + \sum_{n=1}^{\infty}c_{n-1}\,(x-1)^{n+s} = 0$$

$$\sum_{n=0}^{\infty}[2\,(n+s)+1]\,[(n+s)+1]\,c_n\,(x-1)^{n+s} + \sum_{n=1}^{\infty}c_{n-1}\,(x-1)^{n+s} = 0$$

$$(2s+1)\,(s+1)\,c_0 x^s + \sum_{n=1}^{\infty}[(2n+2s+1)\,(n+s+1)\,c_n + c_{n-1}]\,(x-1)^{n+s} = 0.$$

Equating coefficients of like powers of $(x-1)$ gives $(2s+1)\,(s+1) = 0$ for $c_0 \neq 0$, so $s = -1/2$ or $s = -1$, and

$$(2n+2s+1)\,(n+s+1)\,c_n + c_{n-1} = 0, \qquad n = 1,2,3,\cdots,$$

or

$$c_n = \frac{-1}{(2n+2s+1)\,(n+s+1)}c_{n-1}, \qquad n = 1,2,3,\cdots.$$

For $s = -1/2$,

$$c_n = \frac{-1}{(2n)(n+1/2)} c_{n-1} = \frac{-1}{n(2n+1)} c_{n-1}, \qquad n = 1, 2, 3, \cdots.$$

$c_1 = -c_0/(1 \cdot 3)$,
$c_2 = -c_1/(2 \cdot 5) = (-1)^2 c_0/(2 \cdot 1 \cdot 5 \cdot 3)$,
$c_3 = -c_2/(3 \cdot 7) = (-1)^3 c_0/(3 \cdot 2 \cdot 1 \cdot 7 \cdot 5 \cdot 3)$,
$c_n = (-1)^n c_0/[n!(2n+1)(2n-1)\cdots(5)(3)]$, $n = 1, 2, 3, \cdots$. Thus,

$$y_1(x) = c_0 (x-1)^{-1/2} \left[1 + \sum_{n=1}^{\infty} \frac{(-1)^n (x-1)^n}{n!(2n+1)(2n-1)\cdots(5)(3)}\right]$$

is one solution.

For $s = -1$,

$$c_n^* = \frac{-1}{(2n-1)n} c_{n-1}^*, \qquad n = 1, 2, 3, \cdots.$$

$c_1^* = -c_0^*/(1 \cdot 1)$,
$c_2^* = (-1)^2 c_0^*/(3 \cdot 1 \cdot 2 \cdot 1)$,
$c_3^* = (-1)^3 c_0^*/(5 \cdot 3 \cdot 1 \cdot 3 \cdot 2 \cdot 1)$,
$c_n^* = (-1)^n c_0^*/[(2n-1)(2n-3)\cdots(3)(1)n!]$, $n = 1, 2, 3, \cdots$. Thus,

$$y_2(x) = c_0^* (x-1)^{-1} \left[1 + \sum_{n=1}^{\infty} \frac{(-1)^n (x-1)^n}{n!(2n-1)(2n-3)\cdots(3)(1)}\right]$$

is the other solution, and the general solution is

$$y(x) = y_1(x) + y_2(x).$$

3. **(c)** Write the differential equation $x(x+1)^2 y'' - (x^2 + 3x + 2) y' + 9y = 0$ as
$[(x+1) - 1](x+1)^2 y'' - \left[(x+1)^2 + (x+1)\right] y' + 9y = 0$.
$y(x) = \sum_{n=0}^{\infty} c_n (x+1)^{n+s}$,
$y'(x) = \sum_{n=0}^{\infty} (n+s) c_n (x+1)^{n+s-1}$,
$y''(x) = \sum_{n=0}^{\infty} (n+s)(n+s-1) c_n (x+1)^{n+s-2}$.
Substituting these into the differential equation yields

$$[(x+1) - 1](x+1)^2 \sum_{n=0}^{\infty} (n+s)(n+s-1) c_n (x+1)^{n+s-2} +$$

$$-\left[(x+1)^2 + (x+1)\right] \sum_{n=0}^{\infty} (n+s) c_n (x+1)^{n+s-1} + 9 \sum_{n=0}^{\infty} c_n (x+1)^{n+s} = 0$$

$$\sum_{n=0}^{\infty} (n+s)(n+s-1) c_n (x+1)^{n+s+1} - \sum_{n=0}^{\infty} (n+s)(n+s-1) c_n (x+1)^{n+s} +$$

$$-\sum_{n=0}^{\infty} (n+s) c_n (x+1)^{n+s+1} - \sum_{n=0}^{\infty} (n+s) c_n (x+1)^{n+s} + 9 \sum_{n=0}^{\infty} c_n (x+1)^{n+s} = 0$$

$$\sum_{n=0}^{\infty} [(n+s)(n+s-1) - (n+s)] c_n (x+1)^{n+s+1} +$$

$$+\sum_{n=0}^{\infty}[-(n+s)(n+s-1)-(n+s)+9]\,c_n\,(x+1)^{n+s}=0$$

$$\sum_{n=0}^{\infty}(n+s)(n+s-2)\,c_n\,(x+1)^{n+s+1}-\sum_{n=0}^{\infty}(n+s+3)(n+s-3)\,c_n\,(x+1)^{n+s}=0$$

$$\sum_{n=1}^{\infty}(n+s-1)(n+s-3)\,c_{n-1}\,(x+1)^{n+s}-\sum_{n=0}^{\infty}(n+s+3)(n+s-3)\,c_n\,(x+1)^{n+s}=0$$

$$\sum_{n=1}^{\infty}(n+s-1)(n+s-3)\,c_{n-1}\,(x+1)^{n+s}-(s+3)(s-3)\,c_0\,(x+1)^{s}+$$

$$-\sum_{n=1}^{\infty}(n+s+3)(n+s-3)\,c_n\,(x+1)^{n+s}=0$$

$$\sum_{n=1}^{\infty}[(n+s-1)(n+s-3)\,c_{n-1}-(n+s+3)(n+s-3)\,c_n]\,(x+1)^{n+s}+$$

$$-(s+3)(s-3)\,c_0\,(x+1)^{s}=0.$$

Thus, the indicial equation is $(s+3)(s-3)=0$, so $s=3$ and $s=-3$. The recurrence relation is

$$(n+s-1)(n+s-3)\,c_{n-1}-(n+s+3)(n+s-3)\,c_n=0,\qquad n=1,2,3,\cdots,$$

or

$$(n+s+3)(n+s-3)\,c_n=(n+s-1)(n+s-3)\,c_{n-1},\qquad n=1,2,3,\cdots.$$

If $s=-3$, this becomes

$$n(n-6)\,c_n=(n-4)(n-6)\,c_{n-1},\qquad n=1,2,3,\cdots,$$

and we compute the first few coefficients as

$c_1=-3c_0/(1)=-3c_0,$

$c_2=-2c_1/(2)=(-2)(-3)c_0/(2\cdot 1)=3c_0,$

$c_3=(-1)\,c_2/(3)=(-1)(-2)(-3)c_0/(3\cdot 2\cdot 1)=-c_0,$

$c_4=0,$

$c_5=(1)c_4/5=0,$

$(0)c_6=(2)(0)c_5$, so c_6 is arbitrary,

$c_n=(n-4)c_{n-1}/n,\;\; n=7,8,9,\cdots.$

$c_7=3c_6/7.$

$c_8=4c_7/8=(4\cdot 3)c_6/(8\cdot 7)=(4!6!)c_6/(8!2!).$

$c_9=5c_8/9=(5!6!)c_6/(9!2!).$

$c_{10}=6c_9/10=(6!6!)c_6/(10!2!).$

$c_n=(n-4)!6!c_6/[n!2!]$. Thus, the general solution can be written

$$y(x)=c_0\,(x+1)^{-3}\left[1-3\,(x+1)+3\,(x+1)^2-(x+1)^3\right]+c_6\,(x+1)^{-3}\sum_{n=6}^{\infty}\frac{(n-4)!6!}{n!2!}(x$$

3. (e) Write the differential equation $(x-1)^2 y'' - (x-1)(x^2-2x) y' + (x^2-2x) y = 0$ as $(x-1)^2 y'' - \left[(x-1)^3 - (x-1)\right] y' + \left[(x-1)^2 - 1\right] y = 0$.
$y(x) = \sum_{n=0}^{\infty} c_n (x-1)^{n+s}$,
$y'(x) = \sum_{n=0}^{\infty} (n+s) c_n (x-1)^{n+s-1}$,
$y''(x) = \sum_{n=0}^{\infty} (n+s)(n+s-1) c_n (x-1)^{n+s-2}$.
Substituting these into the differential equation yields

$$\sum_{n=0}^{\infty} (n+s)(n+s-1) c_n (x-1)^{n+s} - \sum_{n=0}^{\infty} (n+s) c_n (x-1)^{n+s+2} +$$

$$+ \sum_{n=0}^{\infty} (n+s) c_n (x-1)^{n+s} + \sum_{n=0}^{\infty} c_n (x-1)^{n+s+2} - \sum_{n=0}^{\infty} c_n (x-1)^{n+s} = 0.$$

$$[s(s-1)+s-1] c_0 (x-1)^s + [(s+1)s + (s+1) - 1] c_1 (x-1)^{1+s} +$$

$$+ \sum_{n=0}^{\infty} [(n+3+s)(n+s+1) c_{n+2} - (n+s-1) c_n] (x-1)^{n+s+2} = 0.$$

Thus, the indicial equation is $s^2 - 1 = 0$, so $s = -1$ and $s = 1$. Also $[(s+1)s + (s+1) - 1] c_1 = 0$ and the recurrence relation is

$$(n+3+s)(n+s+1) c_{n+2} - (n+s-1) c_n = 0, \qquad n = 0, 1, 2, \cdots.$$

If $s = 1$, these give $c_1 = 0$ and $(n+4)(n+2) c_{n+2} - n c_n = 0$, so $c_n = 0$, for $n = 2, 3, 4, \cdots$.

Thus, one solution is $y_1(x) = x - 1$. The second solution can be obtained by reduction of order, and we find

$$y_2(x) = (x-1) \int (x-1)^{-2} e^{-\int (x-2)x/(x-1)\,dx} dx = (x-1) \int (x-1)^{-1} e^{-\frac{1}{2}(x-1)^2} dx.$$

We can express $(x-1)^{-1} e^{-\frac{1}{2}(x-1)^2}$ in the form of as power series about $x = 1$, and integrate term by term to find

$$y(x) = c_0 (x-1) + c_0^* (x-1) \left[\ln|x-1| + \sum_{n=1}^{\infty} \frac{(-1)^n}{n! 2^n 2n} (x-1)^{2n} \right].$$

5. (a) If we substitute $y = x^s$ into the left-hand side of the differential equation we obtain $s(s-1)x^{s-1} + sx^{s-1} - 4x^{s+1}$, so the indicial equation is $s(s-1) + s = s^2 = 0$, so $s = 0$ is a double root.

Thus, one solution is of the form $y(x) = \sum_{n=0}^{\infty} c_n x^n$. To find the c_n's we substitute this series into the differential equation to obtain

$$\sum_{n=0}^{\infty} n(n-1) c_n x^{n-1} + \sum_{n=0}^{\infty} n c_n x^{n-1} - 4 \sum_{n=0}^{\infty} c_n x^{n+1} = 0.$$

Thus,

$$\sum_{n=2}^{\infty} n(n-1) c_n x^{n-1} + c_1 + \sum_{n=2}^{\infty} n c_n x^{n-1} - 4 \sum_{n=2}^{\infty} c_{n-2} x^{n-1} = 0,$$

or

$$c_1 + \sum_{n=2}^{\infty} (n^2 c_n - 4 c_{n-2}) x^{n-1} = 0,$$

so $c_1 = 0$ and $n^2 c_n - 4c_{n-2} = 0$, $n = 2, 3, 4, \cdots$. This means that
$c_1 = c_3 = c_5 = c_7 = \cdots = 0$ and
$c_2 = 4c_0/2^2 = c_0$,
$c_4 = 4c_2/4^2 = c_0/4 = c_0/(2!)^2$,
$c_6 = 4c_4/6^2 = c_0/(3!)^2$,
$c_{2n} = c_0/(n!)^2$, $n = 1, 2, 3, \cdots$, and the solution is given by

$$y_1(x) = \sum_{n=0}^{\infty} \frac{1}{(n!)^2} x^{2n}.$$

From Theorem 12.5 we seek a second solution of the form

$$y_2(x) = y_1(x) \ln x + x^0 \sum_{n=1}^{\infty} c_n^* x^n.$$

Substituting this expression into the original differential equation gives

$$(xy_1'' + y_1' - 4xy_1) \ln x + 2y_1' + \sum_{n=2}^{\infty} n(n-1) c_n^* x^{n-1} + \sum_{n=1}^{\infty} nc_n^* x^{n-1} - 4 \sum_{n=1}^{\infty} c_n^* x^{n+1} = 0,$$

or

$$2 \sum_{n=1}^{\infty} \frac{2n}{(n!)^2} x^{2n-1} + \sum_{n=1}^{\infty} (n+1) nc_{n+1}^* x^n + \sum_{n=0}^{\infty} (n+1) c_{n+1}^* x^n - 4 \sum_{n=2}^{\infty} c_{n-1}^* x^n = 0.$$

Combining terms gives

$$4x + \sum_{n=2}^{\infty} \frac{4}{n!(n-1)!} x^{2n-1} + c_1^* + 2c_2^* x + 2c_2^* x + \sum_{n=2}^{\infty} \left[(n+1)^2 c_{n+1}^* - 4c_{n-1}^* \right] x^n = 0.$$

Expand the last series in terms of even and odd terms and rearrange to obtain

$$c_1^* + (4 + 4c_2^*) x + \sum_{n=2}^{\infty} \frac{4}{n!(n-1)!} x^{2n-1} + \sum_{n=1}^{\infty} \left[(2n+1)^2 c_{2n+1}^* - 4c_{2n-1}^* \right] x^{2n} +$$

$$+ \sum_{n=2}^{\infty} \left[(2n)^2 c_{2n}^* - 4c_{2n-2}^* \right] x^{2n-1} = 0.$$

Equating coefficients of like powers of x to zero gives $c_1^* = 0$, $4 + 4c_2^* = 0$,

$$(2n+1)^2 c_{2n+1}^* - 4c_{2n-1}^* = 0, \qquad n = 1, 2, 3, \cdots,$$

$$\frac{4}{n!(n-1)!} + (2n)^2 c_{2n}^* - 4c_{2n-2}^* = 0, \qquad n = 2, 3, 4, \cdots.$$

Thus, $c_1^* = c_3^* = c_5^* = \cdots = 0$, $c_2^* = -1$, and

$$c_{2n}^* = \frac{1}{n^2} c_{2n-2}^* - \frac{1}{n^2 n!(n-1)!}, \qquad n = 2, 3, 4, \cdots,$$

so

$$c_4^* = \frac{1}{2^2} c_2^* - \frac{1}{2^2 2!1!} = \frac{-1}{2^2} - \frac{1}{2^2 2!1!},$$

$$c_6^* = \frac{1}{3^2} c_4^* - \frac{1}{3^2 3!2!} = \frac{-1}{3^2 2^2} - \frac{1}{3^2 2^2 2!1!} - \frac{1}{3^2 3!2!},$$

$$c_8^* = \frac{1}{4^2}c_6^* - \frac{1}{4^2 4!3!} = \frac{-1}{4^2 3^2 2^2} - \frac{1}{4^2 3^2 2^2 2!1!} - \frac{1}{4^2 3^2 3!2!} - \frac{1}{4^2 4!3!},$$

and the general solution is

$$y(x) = c_1 y_1(x) + c_2 \left[y_1(x) \ln x - x^2 + \sum_{n=2}^{\infty} c_{2n}^* x^{2n} \right].$$

Because $x\frac{a_1(x)}{a_2(x)} = 1$ and $x^2\frac{a_0(x)}{a_2(x)} = -4x^2$, $R = \infty$.

5. (c) If we substitute $y = x^s$ into the left-hand side of the differential equation we obtain $s(s-1)x^s + 2sx^s + x^{s+1}$, so the indicial equation is $s(s-1) + 2s = s(s+1) = 0$, so $s = 0$ and $s = -1$.

Thus, one solution is of the form $y(x) = \sum_{n=0}^{\infty} c_n x^n$. To find the c_n's we substitute this series into the differential equation to obtain

$$\sum_{n=0}^{\infty} n(n-1) c_n x^n + \sum_{n=0}^{\infty} 2nc_n x^n + \sum_{n=0}^{\infty} c_n x^{n+1} = 0.$$

Thus,

$$(2c_1 + c_0) x + \sum_{n=2}^{\infty} [n(n+1) c_n + c_{n-1}] x^n = 0.$$

Thus, $c_1 = -c_0/2$ and $n(n+1) c_n + c_{n-1} = 0$, $n = 2, 3, 4, \cdots$. From

$$c_n = -\frac{1}{n(n+1)} c_{n-1}$$

we find

$c_2 = -c_1/(3 \cdot 2) = (-1)^2 c_0/(3 \cdot 2 \cdot 2) = (-1)^2 c_0/(3!2!)$,

$c_3 = -c_2/(4 \cdot 3) = (-1)^3 c_0/(4!3!)$, and

$c_n = (-1)^n c_0/[(n+1)!n!]$ and one solution is given by

$$y_1(x) = c_0 \left[1 - \frac{1}{2}x + \sum_{n=2}^{\infty} \frac{(-1)^n}{(n+1)!n!} x^n \right] = c_0 \left[\sum_{n=0}^{\infty} \frac{(-1)^n}{(n+1)!n!} x^n \right].$$

Trying $s = -1$ gives the same $y_1(x)$, so we set $c_0 = 1$ and seek a second solution of the form

$$y_2(x) = Cy_1(x) \ln x + x^{-1} \sum_{n=0}^{\infty} c_n^* x^n.$$

Substituting this expression into the original differential equation gives

$$C\left(x^2 y_1'' + 2xy_1' + xy_1\right) \ln x + c_0^* + 2xy_1'C + Cy_1 + \sum_{n=3}^{\infty} (n-1)(n-2) c_n^* x^{n-1} +$$

$$+ \sum_{n=2}^{\infty} 2(n-1) c_n^* x^{n-1} + \sum_{n=1}^{\infty} c_n^* x^n = 0,$$

or

$$c_0^* + C(2xy_1' + y_1) + c_1^* x + 2c_2^* x + \sum_{n=3}^{\infty} \left[(n-1) nc_n^* + c_{n-1}^*\right] x^{n-1} = 0.$$

Using the expression for y_1 gives

$$C\left(1 - \frac{3}{2}x + \sum_{n=2}^{\infty} \frac{(-1)^n (2n+1)}{(n+1)!n!} x^n\right) + c_0^* + (c_1^* + 2c_2^*) x + \sum_{n=2}^{\infty} \left[n(n+1) c_{n+1}^* + c_n^*\right] x^n = 0.$$

Equating coefficients of like powers of x to zero gives $C = -c_0^*$, $-3C/2 + c_1^* + 2c_2^* = 0$,

$$C\frac{(-1)^n (2n+1)}{(n+1)!n!} + n(n+1)c_{n+1}^* + c_n^* = 0, \qquad n = 2, 3, 4, \cdots.$$

Because $c_2^* = -c_1^*/2 + 3c_0^*/4$ and

$$c_{n+1}^* = \frac{1}{n(n+1)}\left[-c_n^* + \frac{(-1)^n (2n+1)}{(n+1)!n!}c_0^*\right], \qquad n = 2, 3, 4, \cdots,$$

c_0^* and c_1^* are arbitrary. To simplify the first few terms, we let $c_0^* = 1$ and $c_1^* = 3/2$ so that $c_2^* = 0$, $c_3^* = 5/72$, $c_4^* = -17/1728$ and the general solution is

$$y(x) = C_1 y_1(x) + C_2\left[-y_1(x)\ln x + x^{-1}\left(1 - \frac{3}{2}x + \frac{5}{72}x^3 - \frac{17}{1728}x^4 + \cdots\right)\right]. \qquad R = \infty.$$

5. (e) If we substitute $y = x^s$ into the left-hand side of the differential equation we obtain $s(s-1)x^{s-1} + sx^s - sx^{s-1} - x^s$, so the indicial equation is $s(s-1) - s = s(s-2) = 0$, giving $s = 0$ and $s = 2$. Thus, one solution is of the form $y(x) = \sum_{n=0}^{\infty} c_n x^n$. To find the c_n's we substitute this series into the differential equation to obtain

$$\sum_{n=0}^{\infty} n(n-1)c_n x^{n-1} + \sum_{n=0}^{\infty} nc_n x^n - \sum_{n=0}^{\infty} nc_n x^{n-1} - \sum_{n=0}^{\infty} c_n x^n = 0.$$

Combining terms gives

$$-c_1 - c_0 + \sum_{n=1}^{\infty} [(n+1)(n-1)c_{n+1} + (n-1)c_n]x^n = 0.$$

Thus, $-c_1 - c_0 = 0$ and $(n+1)(n-1)c_{n+1} + (n-1)c_n = 0$, $n = 1, 2, 3, \cdots$. With $n = 1$ the last equation becomes

$2 \cdot 0c_2 = -0c_1$, so c_0 and c_2 are arbitrary, and

$c_1 = -c_0$,

$c_3 = -c_2/3$,

$c_4 = -c_3/4 = (-1)^2 c_2/(4 \cdot 3)$,

$c_5 = -c_4/5 = (-1)^3 c_2/(5 \cdot 4 \cdot 3)$, and

$c_n = (-1)^n 2c_2/n!$. Thus, the general solution is given by

$$y(x) = c_0(1-x) + 2c_2\left[\sum_{n=2}^{\infty} \frac{(-1)^n}{n!}x^n\right].$$

But because

$$e^{-x} = \sum_{n=0}^{\infty} \frac{(-1)^n}{n!}x^n$$

then

$$y(x) = c_0(1-x) + 2c_2\left(e^{-x} + x - 1\right).$$

Note: we could rewrite this solution as

$$y(x) = (c_0 - 2c_2)(1-x) + 2c_2e^{-x} = C_1(1-x) + C_2e^{-x}.$$

Because $x\frac{a_1(x)}{a_2(x)} = x - 1$ and $x^2\frac{a_0(x)}{a_2(x)} = -x$, $R = \infty$, as is obvious from the last form of the answer.

5. (g) If we substitute $y = x^s$ into the left-hand side of the differential equation $x^2y'' + (x^2 - x)\,y' - (x-1)\,y = 0$ we obtain $s(s-1)x^s + (x^2 - x)\,sx^{s-1} - (x-1)\,x^s$, so the indicial equation is $s(s-1) - s + 1 = (s-1)^2 = 0$, so $s = 1$. Thus, one solution is of the form $y(x) = \sum_{n=0}^{\infty} c_n x^{n+1}$. To find the c_n's we substitute this series into the differential equation to obtain

$$\sum_{n=0}^{\infty} n^2 c_n x^{n+1} + \sum_{n=0}^{\infty} nc_n x^{n+2} = 0.$$

Combining terms gives

$$\sum_{n=0}^{\infty} \left[(n+1)^2 c_{n+1} + nc_n\right] x^{n+2} = 0.$$

This requires that $(n+1)^2 c_{n+1} + nc_n = 0$, for $n = 0, 1, 2, \cdots$, so $c_1 = c_2 = c_3 = \cdots = 0$, giving

$y_1(x) = c_0 x$ as one solution. We can obtain the second solution by reduction of order

$$y_2(x) = x \int x^{-2} e^{-\int (x^2 - x)/x^2\, dx}\, dx = x \int x^{-1} e^{-x}\, dx.$$

We can express $x^{-1}e^{-x}$ in the form of a power series about $x = 0$, and integrate term by term to find

$$y(x) = c_0 x + c_0^* \left[x \ln x + \sum_{n=1}^{\infty} \frac{(-1)^n}{nn!} x^{n+1}\right], \qquad R = \infty.$$

5. (i) If we substitute $y = x^s$ into the left-hand side of the differential equation $xy'' + xy' + y = 0$ we obtain $s(s-1)x^{s-1} + sx^s + x^s = 0$, so the indicial equation is $s(s-1) = 0$, giving $s = 0$ and $s = 1$. Thus, with $s = 1$, one solution is of the form $y(x) = \sum_{n=0}^{\infty} c_n x^{n+1}$. To find the c_n's we substitute this series into the differential equation to obtain

$$\sum_{n=0}^{\infty} n\,(n+1)\,c_n x^n + \sum_{n=0}^{\infty} (n+1)\,c_n x^{n+1} + \sum_{n=0}^{\infty} c_n x^{n+1} = 0$$

Combining terms gives

$$\sum_{n=0}^{\infty} \left[(n+1)\,(n+2)\,c_{n+1} + (n+2)\,c_n\right] x^{n+1} = 0.$$

This requires that $(n+1)\,c_{n+1} + c_n = 0$, for $n = 0, 1, 2, \cdots$, so

$$c_{n+1} = \frac{-1}{n+1} c_n = \frac{(-1)^{n+1}}{(n+1)!} c_0,$$

giving $c_n = (-1)^n c_0/n!$. Thus, one solution is $y_1(x) = c_0 \sum_{n=0}^{\infty} (-1)^n x^{n+1}/n! = c_0 x \sum_{n=0}^{\infty} (-1)^n x^n/n! = c_0 x e^{-x}$. Using reduction of order, the second solution is given by

$$y_2(x) = xe^{-x} \int \frac{1}{x^2 e^{-2x}} e^{\int -1\, dx}\, dx = xe^{-x} \int \frac{e^x}{x^2}\, dx.$$

Expanding e^x as $e^x = \sum_{n=0}^{\infty} x^n/n!$ and integrating gives

$$y_2(x) = xe^{-x} \left[-\frac{1}{x} + \ln|x| + \sum_{n=2}^{\infty} \frac{x^{n-1}}{(n-1)\,n!}\right].$$

Expanding e^{-x} and collecting terms gives

$$y_2(x) = \left(x - x^2 + \frac{1}{2}x^3 - \frac{1}{6}x^4 + \cdots\right)\ln|x| + x - \frac{1}{12}x^3 + \frac{5}{36}x^4,$$

and the general solution as $y(x) = y_1(x) + y_2(x)$, $R = \infty$.

7. If we substitute $y = x^s$ into the left-hand side of the differential equation $(x - x^2)\,y'' - 3xy' - y = 0$ we obtain $s(s-1)x^{s-1} - s(s-1)x^s - 3sx^s - x^s = 0$, so the indicial equation is $s(s-1) = 0$, giving $s = 0$ and $s = 1$. Thus, with $s = 1$, one solution is of the form $y(x) = \sum_{n=0}^{\infty} c_n x^{n+1}$. To find the c_n's we substitute this series into the differential equation to obtain

$$\sum_{n=0}^{\infty} n(n+1)\,c_n x^n - \sum_{n=0}^{\infty} n(n+1)\,c_n x^{n+1} - \sum_{n=0}^{\infty} 3(n+1)\,c_n x^{n+1} - \sum_{n=0}^{\infty} c_n x^{n+1} = 0$$

Combining terms gives

$$\sum_{n=0}^{\infty}\left[(n+1)(n+2)\,c_{n+1} - (n+2)^2\,c_n\right]x^{n+1} = 0.$$

This requires that $(n+1)\,c_{n+1} - (n+2)\,c_n = 0$, for $n = 0, 1, 2, \cdots$, so

$$\frac{1}{n+2}c_{n+1} = \frac{1}{n+1}c_n = c_0,$$

giving $c_n = (n+1)\,c_0$. Thus, one solution is $y_1(x) = c_0\sum_{n=0}^{\infty}(n+1)\,x^{n+1} = c_0 x\sum_{n=0}^{\infty} nx^{n-}$ $c_0 x/(1-x)^2$. Using reduction of order, the second solution is given by

$$y_2(x) = \frac{x}{(1-x)^2}\int \frac{(1-x)^4}{x^2}e^{\int 3/(1-x)\,dx}\,dx = \frac{x}{(1-x)^2}\int \frac{1-x}{x^2}\,dx = -\frac{x}{(1-x)^2}\ln x - \frac{1}{(1-}$$

9. If we substitute $y(x) = \sum_{n=0}^{\infty} c_n x^{n-1}$ into $x(x-1)\,y'' + (5x-2)\,y' + 4y = 0$ we find

$$\sum_{n=0}^{\infty}\left[-n(n+1)\,c_{n+1} + (n+1)^2\,c_n\right]x^{n-1} = 0.$$

This requires that $-nc_{n+1} + (n+1)\,c_n = 0$, for $n = 0, 1, 2, \cdots$, so $c_0 = 0$ and

$$\frac{1}{n+1}c_{n+1} = \frac{1}{n}c_n = c_1.$$

Thus, $y(x) = \sum_{n=0}^{\infty} c_n x^{n-1} = c_1\sum_{n=0}^{\infty} nx^{n-1}$, which agrees with Example 12.21.

11. If we substitute $y = x^s$ into the left-hand side of the differential equation $x^2y'' + (x^2 - 3x)\,y' + (-2x + 4)\,y = 0$ we obtain $[s(s-1) - 3s + 4]\,x^s + sx^{s+1} - 2x^{s+1}$, so the indicial equation is $s(s-1) - 3s + 4 = (s-2)^2 = 0$, so $s = 2$ is a double root. Thus, one solution is of the form $y(x) = \sum_{n=0}^{\infty} c_n x^{n+2}$. To find the c_n's we substitute this series into the differential equation to obtain

$$\begin{gathered}\sum_{n=0}^{\infty}(n+2)(n+1)\,c_n x^{n+2} + \sum_{n=0}^{\infty}(n+2)\,c_n x^{n+3} + \\ -\sum_{n=0}^{\infty} 3(n+2)\,c_n x^{n+2} - \sum_{n=0}^{\infty} 2c_n x^{n+3} + \sum_{n=0}^{\infty} 4c_n x^{n+2} = 0\end{gathered}$$

Combining terms gives

$$\sum_{n=0}^{\infty}\left[(n+1)^2\,c_{n+1} + nc_n\right]x^{n+3} = 0$$

This requires that $(n+1)^2 c_{n+1} + nc_n = 0$, for $n = 0, 1, 2, \cdots$, so $c_1 = c_2 = c_3 = \cdots = 0$, giving

$y_1(x) = c_0 x^2$ as one solution. We can obtain the second solution by reduction of order

$$y_2(x) = x^2 \int x^{-4} e^{-\int (x^2 - 3x)/x^2 \, dx} \, dx = x^2 \int x^{-1} e^{-x} \, dx.$$

We can express $x^{-1} e^{-x}$ in the form of a power series about $x = 0$, and integrate term by term to find

$$y_2(x) = x^2 \ln x + x^2 \sum_{n=1}^{\infty} \frac{(-1)^n}{nn!} x^n, \qquad R = \infty.$$

Notes

Notes

Notes

Notes

Notes

Notes

Notes

Notes